The publisher and the University of California Press Foundation gratefully acknowledge the generous support of the Ralph and Shirley Shapiro Endowment Fund in Environmental Studies.

Cane Toad Wars

ORGANISMS AND ENVIRONMENTS
Harry W. Greene, Consulting Editor

Cane Toad Wars

Rick Shine

UNIVERSITY OF CALIFORNIA PRESS

University of California Press, one of the most distinguished university presses in the United States, enriches lives around the world by advancing scholarship in the humanities, social sciences, and natural sciences. Its activities are supported by the UC Press Foundation and by philanthropic contributions from individuals and institutions. For more information, visit www.ucpress.edu.

University of California Press
Oakland, California

© 2018 by The Regents of the University of California

Library of Congress Cataloging-in-Publication Data

Names: Shine, Richard, author.
Title: Cane toad wars / Rick Shine.
Description: Oakland, California : University of California Press, [2018] | Includes bibliographical references and index.
Identifiers: LCCN 2017044544 (print) | LCCN 2017048226 (ebook) | ISBN 9780520967984 (ebook) | ISBN 9780520295100 (paper laminated cloth (plc) : alk. paper)
Subjects: LCSH: Bufo marinus—Australia. | Biological invasions—Australia.
Classification: LCC QL668.E227 (ebook) | LCC QL668.E227 S55 2018 (print) | DDC 597.8/70994—dc23
LC record available at https://lccn.loc.gov/2017044544

Manufactured in the United States of America

26 25 24 23 22 21 20 19 18
10 9 8 7 6 5 4 3 2 1

CONTENTS

Foreword by Harry W. Greene vii
Preface xi

1 · An Ecological Catastrophe 1
2 · How the Cane Toad Came to Australia 13
3 · Arrival of Cane Toads at Fogg Dam 36
4 · How Cane Toads Have Adapted and Dispersed 55
5 · The Impact of Cane Toads on Australian Wildlife 79
6 · How the Ecosystem Has Fought Back 108
7 · Citizens Take On the Toad 130
8 · The Quest for a Way to Control the Toad 155
9 · A New Toolkit for Fighting the Toad 178
10 · Toad Control Moves from the Lab to the Field 203
11 · What We've Learned 228

Acknowledgments 245
Appendix 247
Bibliography and Suggested Reading 251
Index 259

FOREWORD

Cane Toad Wars is the fifteenth installment in the University of California Press's Organisms and Environments series, whose unifying themes are the diversity of life, the ways that living things interact with each other and their surroundings, and the implications of those relationships for science and society. We have sought works that promote unusual and unexpected connections among seemingly disparate subjects, and that are distinguished by the unique talents and perspectives of their authors. Previous volumes spanned topics as diverse as treeshrew behavior and grassland ecology, but none have addressed such an extremely complex conservation dilemma as the one explored herein.

Cane Toad Wars engagingly recounts the most recent phase in Richard Shine's illustrious, career-long obsession with a fascinating continent and its biota. Previously, Rick had mainly studied the natural history of lizards and snakes, but here he turns to the ecological, cultural, and political processes underlying a notorious alien's success story—indeed, his creatures of choice are famously emblematic of human intervention gone haywire. Technically, alien species are those not present in a region prior to modern times; invasive species are aliens whose exploding populations negatively affect native ecosystems and/or human welfare. Averaging roughly two pounds each, Cane Toads were introduced from the Neotropics to the Land Down Under in 1935 to control beetle pests in commercial sugarcane. They are veritable eating and breeding machines and—squat bodies notwithstanding—astonishing long-distance hoppers. Within decades, these drab (some would say grotesque) amphibians had failed at their intended purpose but increased dramatically in population size and distribution, their explosive deployment causing catastrophic declines among some endemic wildlife. In the pantheon of invasive species, Cane Toads were soon superstars.

Rick's story unfolds across the eastern and northern parts of Australia, that long-isolated, mostly arid southern landmass so famous for adaptive radiations of marsupials, of lizard groups only modestly successful elsewhere, and of cobra relatives. What began with its author's concern for the invasive amphibians' potential impact on snakes at his favorite research site turned into a massive, multifaceted attack on the problem. Starting from field observations, taking ecological and evolutionary theory into account, Rick's team designed ambitious experiments; when results were surprising, they revisited theory, headed back into nature, and conducted more experiments. Over the years, he discovered a new form of selection, reached novel conclusions about how best to deal with the toads, and came to admire them. By my reading, these efforts have been successful by virtue of Rick's love of nature and boundless curiosity, but also because of his phenomenal ability to manage people, frame hypotheses for testing, and keep an eye on the matter at hand while thinking outside the box.

This book also reveals how a scientist's research agenda and personal choices can take unexpected turns—from being primarily aimed at answering fundamental questions in ecology and evolution to solving a problem centered on a single species; from framing questions in terms of simplistic, widely accepted concepts to more expansive, nuanced considerations and longer timescales. Now, thanks to years of teamwork—taking data on tens of thousands of tadpoles and toads, getting grants, writing dozens of peer-reviewed publications—some key answers are at hand. Meanwhile, Rick has been honored by Australia's highest awards for scientific achievement, a testament to the importance of studying organisms in the field, careful experimentation, and placing one's work in broader contexts. That *Cane Toad Wars* nonetheless will generate controversy is among its many strongpoints, especially because its author's personal biases are unabashedly on display and because he always circles back to evidence.

For us seasoned professionals, *Cane Toad Wars* passionately synthesizes the capstone endeavors of a renowned biologist—his painstaking research and its yield of far more than the sum of individual projects. For conservation's next generation of practitioners, this volume provides a candid, thoughtful look at how the work gets done and some hints for the path ahead. Lay readers, too, will enjoy Rick's vivid tales from the fieldwork, as well as his lucid explanations and clearly articulated vision of how societies might cope with invasive species. Most profoundly, *Cane Toad Wars* lays out two crucial lessons for everyone concerned about the fate of biological diversity. First, effective conservation is about more than vilifying aliens and saving endangered species—it also must entail realistic adjustments to our notions of the native, the natural, and the wild. Second, conservation outcomes ultimately hinge on conflicting values—themselves shaped by complex and often murky cultural, economic, and political influences—but evidence-based science should play central roles in just how those struggles play out.

Harry W. Greene

PREFACE

Native to Latin America, the Cane Toad became an international traveler through the misguided enthusiasm of sugarcane growers. A gigantic amphibian with a prodigious appetite for bugs seemed like a perfect eradicator of cane-eating insects, so 101 toads were brought to Australia. They were an ecological disaster. After the toads were released in 1935, they swept across the tropics, fatally poisoning millions of native predators in the process. Eventually, in 2005, cane toads arrived at my research site at Fogg Dam in tropical Australia.

For more than twenty years, I had been studying the ecology of snakes on this tropical floodplain. It's a wonderland for any biologist, an amazing natural laboratory—and it was in peril. The toads looked set to destroy an ecosystem that had dominated my research career and that has a special place in my heart.

Many people had tried to stop the toad invasion, but they had failed. Killing toads was futile; the numbers were too great. So, in desperation, I decided to understand toads—not kill them. To switch my research from snakes to toads. That was not an easy decision. Born in Brisbane, I grew up in a world where Cane Toads were loathed as ugly, evil aliens. As a snake ecologist, I had ignored frogs of all kinds. But then Cane Toads took over my research world.

So my snake research took a backseat, and my team geared up to find out what Cane Toads do and what effects they have on native wildlife—and, with luck, to solve some of the problems created by the Cane Toad in Australia. To a significant degree, we succeeded. In the process, I learned about issues that are more general than just toads or just Australia—how environmental calamities play out in the public arena, how people react to them, and how depending on evidence versus intuition affects one's response to conservation crises.

On our rapidly degrading planet, every species is encountering novel challenges. But few face the avalanche of new predicaments encountered by the pioneer individuals at the forefront of a biological invasion. Thus, invaders offer unique opportunities to learn how a species adapts to a changing world. This book tells the story of the Cane Toad, the problems it has created in Australia, and what it has taught us about how to address those problems.

ONE

An Ecological Catastrophe

> A careful observation of Nature will disclose pleasantries of superb irony. She has for instance placed toads close to flowers.
> **HONORÉ DE BALZAC**, *Massimilla Doni*, 1837

This book tells a story of warfare, at several levels and among several combatants. It's a tale of an invasive amphibian that has devastated native wildlife in Australia; of how the ecosystem fought back to get the invader under control; of battles between scientists who championed the the toad's introduction and those who opposed it; of claims and counterclaims regarding toad impact and management, fought out in the public arena by scientists and community toad-busting groups; and of how my research team developed a new arsenal of weapons to control the Cane Toad. Ironically, we stole most of those weapons from the toads themselves, by eavesdropping on tricks they use to kill their competitors—the fiercest battles involve Toad against Toad.

But I was blissfully unaware of all those complexities when I became embroiled in the War of the Toad. All I knew was that the toad invasion was rolling westward across tropical Australia and would soon arrive at my doorstep; that the arrival of Cane Toads spelled doom for many native animals; and that vast energy and effort from other scientists, as well as from the general community, had failed to slow the toad's progress.

At its worst, a biological invasion is a nightmare. Suddenly confronted by a new type of enemy, even abundant and widespread species may be faced with ecological Armageddon. We see it most clearly with emerging diseases: plagues like the Black Death wiped out one-third of Europe's human population in the fourteenth century; the Chytrid Fungus is obliterating frogs on every continent; and Chestnut Blight Fungus kills even century-old trees. But larger invaders wreak havoc as well. All around the world, the arrival of rats, cats, and people has eradicated unique wildlife. And in Australia, Cane Toads are writing a new chapter in the dismal history of invader devastation.

Isolated from the other continents for millions of years, Australia evolved its own unique fauna. Instead of deer, squirrels, and beavers, we have Red Kangaroos, Koalas, and Platypuses. Instead of drab little songbirds, my front garden in Sydney is home to raucous Sulphur-crested Cockatoos and ornate Rainbow Lorikeets. And instead of garter snakes, rattlesnakes, and adders, I find Red-bellied Black Snakes and Diamond Pythons. (Species' common names are capitalized throughout, whereas common names that refer to groups of species are uncapitalized. All common names of species are listed in alphabetical order in the appendix, alongside their scientific names.) Geographically inaccessible to the animal groups that dominate other continents, Australia has been a cradle for the development of life-forms and ecological dynamics found nowhere else. But that uniqueness has a downside: a vulnerability to invasion. If introduced to Australia, animals and plants from North America, South America, Europe, and Asia can blindside the native species. Species within an ecosystem evolve side by side, which gives them an opportunity to adapt to each other. Predators, prey, and parasites are locked in evolutionary "arms races," each finely tuned to the threats and opportunities posed by the other species.

Invasion breaks those rules. It exposes an entire fauna and flora to a type of organism they have never encountered before. That newcomer can be devastating. So it was with the Cane Toad. It has no close relatives in Australia—members of its lineage (the "true toads": Family Bufonidae, or "bufonids") never reached our continent, though they occur widely on others. And early in their long evolutionary history, the bufonids developed a powerful defense against predators: a potent poison. In the toads' native range, local predators adapted to that chemical via a gradual arms race; they can eat toads without dying. But Australian predators never had that evolutionary opportunity. Even a drop of toad poison is deadly.

As a result, native wildlife is being massacred across Australia. That slaughter began when toads arrived in 1935, and the wave of death is moving faster and faster. Nobody paid attention to the carnage in the first few decades after toads began to spread. Most of the Cane Toad's victims were animals that people disliked (like crocodiles and snakes) or that were a threat to poultry (like large lizards and predatory marsupials). But as environmental awareness grew, and unreasoning hatred of reptiles receded, the terrible truth became clear. In 1975, two biologists at the Queensland Museum published a landmark paper drawing attention to the catastrophic mortality of wildlife caused by the Cane Toad invasion.

A TOAD IS A TYPE OF FROG, BUT NOT VICE VERSA

All of the 6,000-plus species of modern frogs and toads are closely related. They form the Anura, one of three main lineages of amphibians (the others are salamanders and the limbless burrowers and swimmers called "caecilians"). All the Anura are called "frogs"—and some of those frogs are toads.

Today's anurans evolved from an ancestor that lived 250 million years ago. Some of its descendants live in the trees—slender-waisted, long-limbed, elfin creatures with big eyes and large toe-pads. Others live in water, with webbed feet that aid their swimming. Most of the ones that live on land are brown (and thus are well camouflaged), with feet better suited to hopping or digging than to clinging. Several evolutionary lineages of frogs have opted for the ground-frog lifestyle, and they all look similar. Trying to identify a small brown ground-frog in the forest at night, with mosquitoes buzzing in your ears and a dim flashlight for illumination, is one of the most frustrating experiences in tropical fieldwork.

But toads are easy to identify; they stand out among the fifty major families of frogs. Toads are toads—in shape, posture, and behavior. Evolution worked out a good design 50 million years ago, and most of the five hundred living species of toads have retained that design. These so-called "true toads" (Family Bufonidae) are a very old group in geological terms, but they haven't changed much. The species from 40 million years ago would be easily recognizable as toads if they came back to life and hopped across your backyard.

The Bufonidae also include some weird creatures that deviate from that prototype, but most modern toads have opted for conservatism in their ecological niche as well as their body shape. Regardless of whether you find them in the rainforests of Brazil or the suburbs of London, these Universal Toads are short, fat, slow-moving ground-dwellers that live around water bodies, eat anything that moves, and protect themselves with a toxic cocktail of chemicals in glands on their shoulders.

Lacking the athletic ability of active mega-leapers like the aptly named "rocket frogs," the climbing ability of the tree frogs, or the underwater acrobatics of aquatic frogs, toads make do by sitting in moist spots and grabbing unlucky insects that come too close. Their sex lives are also staid. Many other types of frogs have complicated reproductive strategies, laying their eggs on land or even incubating the babies inside the father's vocal sac or the mother's uterus or stomach. But the Universal Toad eschews such new-fangled frippery. It lays its eggs—and lots of them—in strings in the nearest pond, and these hatch into small tadpoles that emerge from the water as tiny toadlets—miniature versions of their parents.

The Cane Toad is one of the largest and most toxic amphibians in the world. Photo by Matt Greenlees.

It took another three decades before detailed studies were conducted to measure the impact of Cane Toads on wildlife. In site after site across tropical Australia, that research showed that within a few months of Cane Toads arriving in an area, more than 90 percent of the "top predators" were dead. The rotting bodies of Freshwater Crocodiles floated downstream. The corpses of giant varanid lizards ("goannas") dotted the floodplain. Species once common, like the Northern Quoll and the massive Bluetongue Lizard, could no longer be found. These are apex predators—critically important for ecosystem function. They control the numbers of smaller species. Take the top predators away and everything changes. In North America, the near eradication of Wolves and Mountain Lions was followed by an explosion in deer abundance and massive overbrowsing of vegetation. The ecosystem changed. It is changing in tropical Australia as well, as Cane Toads mow down the top predators, and we still don't know where it will all end.

How could this happen? How could an animal that evolved in the Amazon—in a warm, wet world—survive and flourish in the harsh Australian outback?

From the perspective of its victims, the Cane Toad was the wrong animal in the wrong place at the wrong time.

To explain the toad's success in Australia, we have to go back to its origins in South America. Cane Toads belong to a pioneering group of amphibians—the bufonids—whose transcontinental invasions make Genghis Khan and his Mongol hordes look like introverted homebodies. Toads thrive in most parts of the world because they are ecological generalists. Thirty-five million years ago, ancestral toads embarked on a circumglobal journey that would end with world domination. Beginning in South America, these humble, squat little creatures achieved an extraordinary diaspora from Acapulco to Zanzibar.

Not all toads are created equal. Some kinds of toads were better than others at marching across continents and floating on driftwood over oceans. Across millions of years and the entire circumference of the globe, the pioneering toads were those with land-dwelling adults (so they could move from one pond to the next), large poison glands (that kept them safe from predators), fat-storage organs (to survive the bad times, when food was scarce), opportunistic breeding (so they could make babies whenever the chance arose), and a large number of offspring in every clutch (so at least a few babies survived). Once they had evolved that set of characteristics, the South American toads took off to North America, Africa, Europe, and Asia.

So, toads were Great Invaders long before a supersized species evolved 20 million years ago in the Amazon Basin, at the margins of the great rainforests. This giant species—the Cane Toad—was destined to become the ultimate world traveler, eclipsing all of its relatives. Nonetheless, it was just building on a long family tradition, established eons before prehumans evolved in the African savannas 2 million years ago. And the final missing element in the Bufonidae's global conquest fell into place in 1935, when thirty-three-year-old Reg Mungomery collected 101 Cane Toads in Hawaii and brought them back to Australia. Reg brought toads to the only toad-habitable continent that they hadn't reached on their own.

The sugarcane growers who invited the Cane Toad to Australia wanted an animal that would eat as many insect pests as possible. And they succeeded: given the opportunity, Cane Toads eat hundreds of prey items in a single night. When you pick up a well-fed Cane Toad, it crackles in your hand from insect exoskeletons rubbing together in the toad's stomach. The cane farmers didn't realize that they had also chosen an awesome invader. Cane Toads inherited a capacity for long-distance dispersal, powerful toxins, a flexible lifestyle, and a prodigious reproductive output. They are the living embodiment of the characteristics that enabled ancestral bufonids to spread around the planet, primarily because they are among the largest toads. Being bigger gives you longer legs (increasing your mobility), bigger poison glands (increasing your toxicity), an ability to eat large as well as small prey (allowing you to adapt your diet to local conditions), and a capacity to produce vast numbers of eggs (there is a lot of room for eggs inside a female Cane Toad). If you tried to design an Amphibian Invasion Machine, you couldn't do much better.

Five characteristics of Cane Toads have made them extraordinarily good invaders. First, they are tough. Their large body size buffers them against unfavorable conditions. The bigger you are, the longer it takes to dry out or overheat; and you can travel a long way to find shelter. And toads tolerate extremes of temperature, salinity, and acidity that would kill a lesser amphibian.

Second, they are flexible. Not in physique, but in behavior and ecology. Toads exploit every opportunity. If a wildfire burns through the forest, Cane Toads flock to the newly formed open clearings to feed. If someone leaves a tap dripping, the toads find it. Take a harsh, inhospitable landscape, but throw in a moist patch with a few bugs—perhaps a pile of poo from a passing cow—and watch the ultimate opportunist hop over to top up his water level, and have dinner as well. As a result of that flexibility, Cane Toads thrive in disturbed areas around towns and cities. The human race has helped the Cane Toad become one of the most abundant amphibians on Earth.

The Cane Toad's flexibility extends to its diet: a toad will eat just about anything it can fit into its mouth. Beetles, bugs, cockroaches, and ants are the staple fare, but Cane Toads will tackle even well-defended creatures like bees, ignoring the powerful sting. After bringing Cane Toads to Hawaii in 1934, Cyril Pemberton wrote that "a toad will, without hesitation, gulp down a bee . . . and apparently suffer no discomfort. After swallowing such a fiery creature as a carpenter bee, Bufo was observed to execute a few abdominal motions suggestive of the Hawaiian 'hula dance.'" Insects dominate the menu for most kinds of frogs, so toads resemble other amphibians in this respect. But frogs won't seize an item unless it moves, whereas toads take nonmoving things also—even bizarre "prey" like lighted cigarette butts, human feces, and rotting vegetables. In Hawaiian botanical gardens, toads died after eating the petals falling from strychnine trees.

Nonetheless, Cane Toads aren't simple-minded gluttons. They use sophisticated tricks to get that evening meal. For example, suburban toads soon learn the location of productive foraging spots, like Fido's food bowl. Beneath bug-

THE BIGGEST OF THE BIGGEST

The biggest Cane Toad ever measured was 38 centimeters (15 inches) in length and weighed 2.8 kilograms (6 pounds). This spoiled pet, named "Prinsen," lived in Sweden. In Swedish, *prinsen* means *prince*—but female toads grow larger than males, so Prinsen may have been female. Regardless, that's a lot of toad. A gentleman from Proserpine in Queensland contacted the Guinness Book of World Records in 1975, claiming that he had an even bigger toad, Gerty, who had grown to 3 kilograms on a diet of beer. Sadly, Gerty passed away before the claim could be confirmed. The biggest toad I've ever seen myself was Bette Davis (named for her beautiful eyes), who lived at the Queensland Museum in the 1980s. She weighed nearly 2 kilograms (4.4 pounds) and was 24 centimeters (9.4 inches) long, about the size of a dinner plate.

The most famous large Cane Toad was Dairy Queen, star of the documentary movie *Cane Toads: An Unnatural History*. Dairy Queen was the prized pet of young Monica. Displaying the nonchalance typical of Cane Toads, Dairy Queen allowed herself to be dressed in frilly frocks and cuddled unmercifully.

In recent years, community-group toad musters have turned up occasional giants also. In 2007, for example, the Northern Territory's "Toadzilla" was proudly introduced to the international media as "being the size of a small dog." Despite the hype, this impressive anuran was only 20.5 centimeters (8 inches) long and weighed 840 grams (less than 2 pounds). By comparison, my West Highland white terriers (Bob and Miss Adelaide) are each about 60 centimeters (24 inches) long and weigh about 10 kilograms (22 pounds). Either someone is stretching the truth, or the dogs in Darwin are the size of guinea pigs. Nonetheless, the world's media uncritically reported the story of the toad that was the size of a small dog.

attracting streetlights, toads space themselves out so that each has enough room around it from which to pick up a fluttering bug. Even in natural areas without artificial lights, Cane Toads in Australia use a similar tactic, an evolutionary precursor to their use of neon lights. The smooth white trunks of gum trees reflect the moonlight, attracting bugs. Cane Toads gather around such a tree, unmoving, staring intently at the gleaming trunk. It looks like a solemn Druid ceremony, the religious observance of a tree-worshipping toad clan.

Unusual among amphibians, Cane Toads will eat nonmoving prey. This toad is eating a road-killed snake. Photo by Crystal Kelehear.

The third factor that has made the Cane Toad such a devastatingly successful invader is its mobility. No other amphibian on the planet has dispersed as far and as fast as the Cane Toad in Australia. Radio-tracked Cane Toads sometimes move more than 2 kilometers (1.2 miles) in a single night. Most amphibians don't move that far in their lifetimes.

Fourth on my list of Toad Tactics is poison: potent toxins that spell death for almost any predator foolish enough to launch an attack. The eggs are full of poison, as are the young tadpoles. And as soon as they transform into miniature toads, the juveniles grow parotoid (shoulder) glands to manufacture their own deadly arsenal. And they are flexible in how much they invest: a tadpole that has been stressed transforms into a young toad with bigger-than-average parotoid glands. It's worth investing in weapons if you live in dangerous times.

Lastly, the Cane Toad's fifth key to success is its incredible reproductive rate. A well-fed female toad can produce two clutches a year, each containing up to 40,000 eggs. As a result, a few pioneer toads can rapidly populate the area with thousands of progeny.

In short, millions of years of evolution at the margins of the South American rainforest created a supersized, toxic, reproductively prolific, mobile, behaviorally flexible amphibian. A creature that could survive the

HOW DEADLY ARE CANE TOADS?

Adult Cane Toads have larger parotoid (shoulder) glands than most other toads—about the size of a walnut. Each gland contains about fifty goblet-shaped lobules (where the poison is stored), connected to the outside by separate ducts. The poison flows through those ducts onto the toad's skin. A threatened toad may ooze poison from its ducts, but the full dose comes out only when the toad is seized by a predator, putting direct pressure on the glands. Some toad species actually shoot the toxin out at aggressors, so we should be grateful that Cane Toads are reluctant to retaliate.

The toad's poisons are a complex mixture of chemicals. They go by many names, broadly called "bufogenins" and "bufotoxins," with different molecules having different biological functions. Our chemical collaborator Rob Capon, at Queensland University, has identified an extraordinary range of substances within this lethal cocktail of milky fluid in the toad's shoulder glands. It's potent stuff. Supposedly, the drug-induced stupor of the Haitian zombies was induced by a mix of Cane Toad toxin and pufferfish poison.

The Campa tribe in Peru (in the upper Amazon) eat Cane Toads, after removing the skin and shoulder glands. But this is Bufonid Roulette unless you know what you're doing. Several people have died from eating Cane Toads, even in the toads' native range, where you expect people to know better. One young boy in a small village in the Peruvian Amazon mistook toad eggs for (edible) frog eggs and made a stew for his family. His mother and his sister were dead within hours.

Most other human deaths from Cane Toad poison have happened at the toad invasion front. If frogs and their eggs are a local delicacy, some adventurous soul may decide to try out the flavor of the big, warty amphibians that just arrived. Rural people who gather their food from the wilderness are at the most risk. The chief of detectives in Iloilo Province on the Island of Panay in the Philippines died after eating three toads that he had mistaken for native frogs. A 1943 newspaper article on Cane Toads in the Australian cane fields commented brutally that "southern European sugar farmers who have tried to eat the giant toad for variety in their menu have become ill."

Most large biological molecules are broken down by heat, but bufotoxins are as tough as the toads that produce them. I learned this by studying road-killed Cane Toads in the Northern Territory. Toads love to sit on roads, and farmers love to flatten them when they drive by—so there is no shortage of dead toads to study. The intense heat means that a toad run over in the dry season soon turns into a pancake—flat (from being repeatedly squashed by trucks), leathery, with all of the moisture sucked out of it. Children use dried toads as Frisbees, and a friend of mine once picked one up, stuck a label on it with my name and address, affixed a stamp, and popped it in the mail. It arrived without any problem, because it was about the size and thickness of a normal letter.

Heating up to 50°C (122°F) in the direct sun every day, for month after month, these parched bodies are scattered along roads throughout tropical Australia. Scavengers ignore them because they are too tough and sun-hardened. But when we put some of these Toad Pancakes into water, they swelled up again and milky poison oozed out of the long-dried shoulder glands. Small aquatic creatures like fishes and native tadpoles avoided it; and if they couldn't get away from it, they died. The poison was still active even after baking in the sun every day for months. If you kill a Cane Toad, dispose of that carcass with care!

tough times and exploit the opportunities that arose when humans cleared the land to build farms and cities. The Cane Toad had already colonized a huge area of South and Central America, but it needed help from humans to take up that challenge in far-flung lands. By giving it that help, we doomed millions of Australian native animals to an agonizing death.

The massacre of native wildlife by invasive toads isn't just an intellectual issue. Death By Toad is quick, but it isn't pretty. And so, as soon as people understood what was going on, the wave of toads spreading across Australia was accompanied by a wave of revulsion. If you enjoy wildlife around your home, and recognize the local lizards individually, it's heart-rending to find those animals lying dead in your backyard.

So, Australians set out to stop the toad invasion. Somehow, anyhow. They killed toads whenever they saw them. They designed and deployed traps. They formed community groups to eradicate toads. They organized toad-busting days. They funded scientists to find new ways to eradicate the hated amphibians. Nothing worked. The toads kept coming. Even though people

A bucketful of Cane Toads gathered by volunteer "toad-busters" from a small park near Darwin. Photo by Terri Shine.

went out every night to destroy the feral amphibians, toads soon outnumbered native frogs. In their frustration, Australians developed a passionate hatred of Cane Toads.

With little reliable information about the Cane Toad or its impacts, myths began to spread: Cane Toads are aggressive, attacking people on sight. Toads grow big enough to swallow your pet dog, and perhaps even your baby. The post-toad-invasion world will be a wasteland full of the corpses of rotting Koalas and Red Kangaroos. If your car won't start in the morning, it's because a Cane Toad has crawled up into the exhaust pipe. Toads poison the drinking water. Your school-age children will abandon their studies to smoke dried toad skins.

As the wave of toads spread across tropical Australia, communities at the invasion front took up the fight. The "toad-busters" were zealous and saw the battle as too important to ignore. But within a few years of toads arriving, the futility of killing them was apparent. The amphibian tsunami kept coming.

And that's where my own involvement with the Cane Toad began. I was faced with a unique opportunity, as the alien amphibians marched toward the site where I had been researching reptiles for twenty years. I could study the enemy on a battlefield that I already knew very well. It was a chance to find out what was really going on during a biological invasion.

I was naive when I began my research on Cane Toads. In truth, I didn't know what I was getting into. I saw Cane Toads as an ecological issue, not a political one. Being a scientist, I saw my role as finding out what toads were doing and how we could stop them from doing it. I didn't realize that Cane Toads were also important to other groups—including community-based "environmental" organizations and politicians at every level, from the local to the federal tier. Or how much passion could be aroused when the conversation turned to Cane Toad control.

Why have Cane Toads become such an iconic animal in Australia? No other amphibian, anywhere in the world, has such a high public profile. Surveys of the Australian public rank Cane Toads as a greater threat to native biodiversity than any other invasive species (including animals like Rabbits and Foxes, which have had far worse ecological impacts). And why do nine out of ten Australians rank Cane Toads as among the worst feral pests? The fundamental answer is simple: Australians love to hate Cane Toads.

The onslaught of these marauding giant amphibians was a "perfect storm" politically as well as ecologically. Public concern about toad impacts created opportunities for aspiring politicians, whose exaggerated rhetoric fueled people's fears. A wave of anti-toad hysteria swept across tropical Australia, in advance of the trespassing Cane Toads.

It's been a steep learning curve. Since I made my leap from snake research into toad research fifteen years ago, I've learned a lot about environmental politics as well as about Cane Toads. The toads have educated me not only in ecological and evolutionary processes, but also in terms of the bizarre circus that plays out at the intersection between science and wider society.

TWO

How the Cane Toad Came to Australia

Bufo marinus, the gentleman in question, is just like any other toad, except that he is far too large.

DINAH STEWART, *The Argus,* Melbourne, May 11, 1951

We don't know much about the Cane Toad's "natural" life at the edge of the South American rainforest. Although it has been intensively studied in Australia, the Cane Toad has been ignored in its native range. Toads aren't a problem there, so why study them? We don't even know how many kinds of giant toads bearing the name "Cane Toad" are wandering around South and Central America. Genetic data hint at four "Cane Toad" species, all of them historically called *Bufo marinus*. For example, populations of "Cane Toads" from east and west of the Andes have been separated for millions of years and have evolved into different species. Only one of those species can be given the name *marinus*. In 2016, scientists finally resurrected an old name *(Bufo horribilis)* for the west-of-Andes species, but other lineages within the group still don't have official names.

Does this mean that the Cane Toad in Australia is a species that has not yet been described by science? That would be ironic, given the Cane Toad's high public profile, but no—the "type specimen" (the original individual from which the Swedish scientist Linnaeus gave the Cane Toad its scientific name) came from French Guiana. So did the ancestors of the toads whose progeny were brought to Australia 177 years later, which means that the animals we call "Cane Toads" in Australia are the same species as the one that Linnaeus named *Rana marina* in 1758.

Our ignorance about Cane Toads in their native range extends to their ecology. For example, Cane Toads are denizens of the towns, not the forests, but how did they evolve that preference? Humans didn't exist when Cane Toads first appear in the fossil record, so there were no towns for the giant

THE NAME OF THE TOAD

Before 1935, Cane Toads were called "Marine Toads" or "Giant Toads." The name "Cane Toad" didn't emerge until after these huge amphibians were brought to Australia to save the sugarcane crop. Why change the animal's name? There was nothing new about the idea of using toads for bug control. In nineteenth-century Europe, toads were released in gardens to reduce insect numbers. Indeed, Cane Toads themselves were used as bug-controllers in sugarcane plantations in Puerto Rico and Hawaii long before they were brought to Australia. But *Bufo marinus* wasn't a high-profile animal until its Australian adventure.

The switch in names was driven by interstate politics. "Cane Toad" took over from about 1949, as the toads spread out from the Queensland coast. Australians encountered warty aliens in their backyards, and they weren't happy. Cane growers were the villains who brought in the toads, so the outraged public from New South Wales identified the culprits by calling the invaders "Queensland Cane Toads." But Aussies prefer brevity, so "Cane Toads" they became.

Bizarrely, the toad's iconic status in Australia has now spread the name "Cane Toad" worldwide. When I visited Costa Rica (within the Cane Toad's native range), a local biologist assured me that Cane Toads were an invasive species there. They aren't; but the crimes of Cane Toads in Australia have blackened their name even in their original home, changing the animal's common name. If ever a species needed to hire a public relations consultant, it's the Cane Toad.

In science, every species is given a Latin name as well as its common name. The custom dates back to Swedish scientist Linnaeus (1707–78), who invented the concept of double-barreled scientific names. The first part is the genus (the group to which the organism belongs) and the second part is the unique species-identifier. Linnaeus himself named the Cane Toad (based on specimens brought from South America to Europe as curiosities) as *Rana marina*. He chose that name because he put all frogs and toads into a single, catchall genus, *Rana;* and this giant toad supposedly could tolerate saltwater (thus *marina*). Linnaeus got this idea about salt tolerance from Dutch zoologist Albertus Seba, who wrote in 1734 that the Cane Toad "is free to live on land and in the sea." It wasn't the last mistake that scientists would make about Cane Toads.

A scientific name is meant to ensure that a species is called the same thing no matter where or when it is studied. The popularity of common names (like "Giant Toad" and "Cane Toad") may wax and wane, but the Latin name remains stable. Ideally, at least. Cane Toads were moved from *Rana* to *Bufo* when the toads got their own genus, and thereafter were called *Bufo marinus* for centuries. But as more information became available, we no longer had to keep all of the toads in a single, large genus *(Bufo)*. Powerful new DNA-sequence-based methods began to reveal subgroups within the genus *Bufo*.

Researchers divided *Bufo* into several subgenera in 2006. The Cane Toad was transferred to the smaller genus *Chaunus,* together with other South and Central American toads. So, as I published our scientific studies, I stopped calling the Cane Toad *Bufo marinus* and started calling it *Chaunus marinus*. But then, only a year later, more DNA work showed that the Cane Toad didn't belong in *Chaunus* after all. Another name *(Rhinella)* was more accurate. So, having only just shifted from *Bufo* to *Chaunus,* the Cane Toad moved to *Rhinella.* To add insult to injury, the rules of Latin meant that the species name had to change also—from *marinus* to *marina.* This left me in the embarrassing position of having published a series of scientific papers in which I variously called the Cane Toad *Bufo marinus, Chaunus marinus,* or *Rhinella marina.* So much for the "stability" of scientific names! And now, sadly, our moniker "Team *Bufo*" is looking out of date (but "Team *Rhinella*" sounds too metrosexual to me). And scientific fashions change. Some die-hard conservatives still reckon the Cane Toad belongs within *Bufo,* so I'm happy to drift along using both names. If you think this really matters, you probably need to get out of the house more often.

amphibians to exploit. Where did toads live before people created disturbed habitat for them? The answer must lie in "natural" disturbances—areas of open ground that are rich in nutrients and water, and degraded by species other than *Homo sapiens*. That degradation can be caused by climatic events (like cyclones and floods) or by "ecosystem engineer" species that trash the world around them. For example, the bare ground beneath communal bird roosts is covered with bird poo, regurgitated food, and decaying nestlings. Insects flock to the smelly supermarket, creating a toad smorgasbord. Perhaps this niche enabled the Cane Toad's early ancestors to evolve into flexible opportunists, able to travel long distances and key in on places where disturbance provides a free meal. When people arrived a few million years later, and began clearing large tracts of land, we turned the world into a Fun Park for Cane Toads.

The most widespread ecological disturbance is agriculture, for which forests are cut down so that crops can be planted and irrigated. Insect pests thrive in commercial plantations, creating opportunities for hungry toads. Agriculture is a battle between people and insects; and in the days before modern insecticides, the insects often won. As global commerce spread in the 1800s, one of the most popular crops was sugarcane (native to New Guinea, but soon grown all over the tropics). Unfortunately for the cane growers, native insects developed a sweet tooth. Sugar yields were jeopardized when grubs (beetle larvae) attacked the roots of the young cane. Many a cane grower, sitting on the balcony on a balmy tropical night, sipping a mint julep, must have pondered how to reduce the impacts of ravenous beetles.

The cane growers of Puerto Rico looked for a scientific solution by setting up the "Insular Experiment Station" at Río Piedras. And before long, the answer to their prayers hopped into sight. In 1923 Menendez Ramos, the station's director, was waiting on the dock in Kingston, Jamaica, for a boat to take him home. Cane Toads were common in Jamaica, having been brought across from mainland South America (apparently via Barbados and Martinique) to take on the insect pests that were cutting into the plantation owners' profits. Toads were gorging themselves on beetles under the dock's lights, and on impulse Ramos scooped up forty of them and took them back with him, to release into his Puerto Rican plantations. That simple action changed the course of history. The toads flourished in their new home, and the concept of using Cane Toads to control sugarcane pests began to spread. The idea received a massive boost a few years later, when the Fourth Congress of the International Society for Sugar Cane Technologists was held in Puerto Rico in 1932. Cane growers from all over the world heard about the New Amphibian Savior. Before long, Cane Toads were being sent to sugarcane and sweet potato plantations everywhere. And among those grateful recipient nations was Australia.

How was that decision made? How did a toad from South America find its way to the other side of the planet, 19,000 kilometers (12,000 miles) away? Fortunately, we have hard evidence of every step along the path. Historian Nigel Turvey has written a wonderful book, *Cane Toads: A Tale of Sugar, Politics and Flawed Science,* based on his firsthand examination of original records.

Commercial sugarcane was big business in the 1930s, and the conference in Puerto Rico brought together many luminaries of the sugarcane world. The first toad devotee was Raquel Dexter, who dissected Cane Toads from

Puerto Rican plantations to see what they had been eating. Having discovered that the toads sometimes ate the beetles that consume cane, she concluded that Cane Toads were a plantation owner's best friend. It was poor science. To assess the impact of Cane Toads on sugar production, we would need to compare yields of plantations with and without toads. Dexter's data showed only that Cane Toads eat beetles. And, to be candid, I'm not a hundred percent convinced of Dexter's expertise in amphibian biology. In her paper, she mentioned that toads spawn in a foamy nest. They don't.

The second major player at the Puerto Rican conference was Cyril Pemberton, the head of sugarcane science in Hawaii. Pemberton was a passionate advocate for biological control and brought many overseas insects to Hawaii "to aid commercial agriculture." He enthusiastically added Cane Toads to that list. Indeed, he liked the animals so much that he kept several as pets from the time they were tadpoles until they had all died of old age fifteen years later. He would later play a starring role in the toad's introduction to Australia, bringing him into open conflict with a man who used to be a close friend. It wouldn't be the last time that Cane Toads destroyed a friendship, as shown by the vitriol of Cane Toad politics in Australia.

Enthused by stories of beetle control in Puerto Rican plantations, Pemberton collected thirty-four Cane Toads while he was at the conference and shipped them home. He carried more toads with him when he returned to Hawaii. A grand total of 166 toads (or 153, depending on which report you believe) had left the Caribbean and traveled halfway around the world. All but five of them made it safely to their new home. It was an astonishing range expansion for a giant amphibian from South America, but it was to pale into insignificance only a few years later, when a young Australian scientist took 101 descendants of those Hawaiian immigrants on yet another journey.

The Australian connection at the 1932 conference in Puerto Rico was Arthur Bell. An agricultural scientist from Queensland, Bell was under political pressure. Beetles were eating Australian cane growers out of business. Many remedies had been tried, including a traveling cage full of chickens that could be released on freshly plowed fields to scratch about for grubs. In other places, the pests were controlled by arsenic powder, insecticides, and sticks of dynamite. Nothing helped. The beetles flourished, the sugarcane withered, and the entrepreneurs scowled.

Like much of the world, Queensland was struggling financially after the Great Depression. Government expenditure was cut to the bone. Nonetheless, the political power of Big Sugar encouraged the Queensland government to approve a princely sum (about half his annual salary) to fund Bell's travel to the conference in Puerto Rico. The decision was kept confidential to avoid howls of outrage from the public at this overseas junket to a tropical island.

To justify his trip, Bell desperately needed a way to combat the Grayback Beetles and Frenchie Beetles that were the cane growers' nemesis. The great Cyril Pemberton, sugar scientist extraordinaire, extolled the Cane Toad's virtues as a biological control agent. Bell was taken with the idea. After he returned to Australia, he handed responsibility for the toad project to a newly appointed researcher, Reg Mungomery. Reg played a central role in the toad debacle, so it's worth looking at his background briefly before we move on with the main story.

Reg was a local lad, born in 1901 in Childers, Queensland. His father was a blacksmith. Young Reg was brilliant. He topped his high school examinations in 1915 and won a gold medal for academic ability two years later at Maryborough Grammar School. In 1920 he scored the Browne memorial medal from the Charters Towers School of Mines. Reg majored in metallurgical engineering, chemistry, and mining surveying, but his heart was in biology. After a stint on the family farm, he was employed by the Prickly Pear Control Board in Rockhampton. Like the Cane Toad, the Prickly Pear is an alien species—a cactus from Brazil that took over vast areas in Australia. The Prickly Pear team's research was (and remains) one of the most successful programs in the history of biocontrol. The pivotal discovery was that a specialist moth (the Cactus Moth) consumes the plant in its native range—and doesn't eat anything else. When the moth was released in Australia, the population of Prickly Pear nosedived. This triumph encouraged the introduction of foreign organisms to solve agricultural problems.

Reg only lasted a year with the Prickly Pear Control Board before moving on to a mining job. And just a year later, in 1924, he joined the organization that would employ him for the rest of his career. The entomology division of the Queensland Bureau of Sugar Experiment Stations must have been heaven on earth for a young Queenslander with a passion for using biology to solve agricultural problems.

Initially, however, Reg resisted the idea of using Cane Toads for biocontrol. In a scientific paper written only a year before he brought Cane Toads to Queensland, Reg pointed out that such attempts often backfire, with the solution proving to be worse than the predicament it was meant to solve. It's worth reading what he said, if only to wonder why he changed his mind a few months later. Writing in the *Cane Grower's Quarterly Bulletin* in 1934, Reg opined,

Brilliant economic successes do not present the whole picture of biological control of insect pests. In this phase of endeavor, perhaps more than any other, the path to successful achievement is strewn with the remains of optimistic attempts which ended in abject failure. Biological control does not consist in rushing off to a foreign country, bringing back a number of parasites, and letting them loose upon the unsuspecting pest; to ensure the success of biological forms of control a whole complex of factors must be interdependently favorable. Such a project is not to be embarked upon lightheartedly, but only after the most mature consideration, since a false step may have disastrous economic consequences through the upsetting of the whole biological balance.

Young Reg initially fought against his boss's plan to use Cane Toads as a path to political salvation. Reg pointed out that the beetles spent very little time on the ground, so the toads wouldn't be able to get at them. The local frogs ate beetles, and they hadn't controlled beetle numbers. Why would toads be any different? Reg was right, and he should have stuck to his guns. Somewhere along the line, though, he changed his mind; the offer of a trip to Hawaii may have helped. The Queensland government funded his journey to collect toads in the cane fields of Oahu.

So, Reg Mungomery headed off on his great adventure. Heady stuff for a young man from outback Queensland, at a time when overseas travel was a luxury. Although now an enthusiastic convert to the idea of importing Cane Toads, Reg received shocking news soon after he reached Hawaii. A two-year-old child had died after eating part of a Cane Toad for dinner. It was the first inkling of the toad's terrible toxicity, but Reg decided to cover it all up. He suggested to Bell that

> it would be a good plan to say nothing of this until I get the toads to Australia and then we can sound a note of warning through the press and the "Quarterly Bulletin" that they are not the edible variety of frogs, but that these toads possess certain poison sacks which render them extremely dangerous if they are eaten.

Reg ignored this golden opportunity to reconsider the plan, pressing on regardless.

Despite falling ill (and having his appendix removed as a result), Reg managed to collect 102 toads from the suburbs of Honolulu. Packed in suitcases with moist wood shavings, the toads set off on the long trip back to Australia. They arrived in Sydney on a steamer, the *Mariposa,* on Monday, June 17, 1935. One of the toads died during the journey, but the 101 survivors received a

When I visited Reg Mungomery's old office at the Sugar Experiment Station in June 2008, his filing cabinet contained hundreds of glass negatives of photographs taken of the toad project. This negative shows the enclosure in which the original Hawaiian toads and their progeny were kept. Photo by Terri Shine.

warm welcome at the Sugar Experiment Station in tropical Queensland, just outside Cairns. While Reg was away catching toads, his colleague James Buzacott had constructed an elegant enclosure. Complete with a pond and Water Hyacinths (another ecologically damaging invader from the Amazon Basin), this dome-shaped Toad Hilton, 3 meters (10 feet) in diameter, was covered with mesh to contain the immigrant amphibians.

The arrival of the toads attracted great fanfare. On June 21, 1935, the Sydney newspapers triumphantly declared "Giant Toads imported to combat pests on canefields. War on insect pests is to be carried out by giant American toads." Under the headline "Bufo's Hearty Appetite Worth Many Pounds to Growers," a Brisbane newspaper predicted that "the sugar industry one day will erect a statue to Bufo marinus." Those words were prophetic. The town center of Sarina, in the cane-growing heartland of North Queensland, now features a giant Cane Toad statue.

Nothing was too much trouble for the savior of the cane crop. The Sugar Experiment Station in Meringa had been open for only a few months, and the Cane Toads were the scientists' first major project. Before long, the pampered amphibians did exactly what Mungomery and Bell had hoped. They bred. The Hawaiians had failed to breed their own toads in captivity, after shipping them in from Puerto Rico, and finally had resorted to releasing the toads in the cane fields. But within a few weeks of arrival, the Australians had achieved this feat. Smugly triumphant grins all around for the Aussies, but this breeding success says more about toad biology than about the loving reception the toads received at Meringa. It takes more than three weeks for a toad to yolk up a clutch of eggs, so Reg must have captured at least one female toad that was ready to breed. After a couple of weeks at sea for those eggs to ripen, the poor female was bursting by the time she encountered the pond at Meringa.

A few years ago, I made a pilgrimage to Reg Mungomery's old office. The research station still exists, and its staff is still trying to stop beetles from eating sugarcane. I was struck by how many photographs the early toad researchers had taken of their charges. Box after box of glass negatives show the excitement of that first breeding success; few newborn human babies have been photographed as often as that first crop of Cane Toads: eggs, early-stage tadpoles, late-stage tadpoles, juveniles. The tide would turn eventually, and Cane Toads would become as popular as dog droppings on the sofa. But in those first golden months, the jubilation was tangible.

The output of eggs was far beyond the sugar scientists' dreams. In their first year after arrival, thirty-seven female toads laid more than 1.5 million eggs. As soon as the first clutch of eggs had passed through the tadpole stage, Reg pedaled off on the research station's pushbike to release bags of toadlets into nearby sites like the Little Mulgrave River. Cane Toads were at liberty in a new land. By late 1935, the genie was out of the bottle.

Newspaper stories of the time exude a robust confidence. The beetles were doomed. Reg wrote that "we confidently expect that [the toad's] presence here will cause no embarrassment, but that it will contribute some material benefit by its incessant attack on the many pests which confront the grower." And "Flattering reports were also being received from some of the housewives, who asserted that the rather disquieting presence of cockroaches in their houses had in many cases been greatly reduced through the activities of the toad." But even as sugar scientists extolled the virtues of Cane Toads, trouble loomed on the horizon: a spirited attempt to block the further release of toads.

The lone voice raised against toad importation was that of Walter Froggatt, an eminent scientist who pointed out (rightly) that Cane Toads might cause more problems than they solved. Froggatt was a retired New South Wales government entomologist, and at least part of the hostility between the sugar scientists and Froggatt seems to have involved interstate rivalry between the Queensland toad importers and their New South Wales toad opponent. That same rivalry persists to this day, notably in football matches. Cane Toads were simply one more issue to divide the northern versus southern Australian states.

Ironically, Walter Froggatt learned of the toads' release from one of its great supporters—his friend Cyril Pemberton, the Hawaiian scientist. Pemberton stayed at Froggatt's house in Sydney after attending a sugarcane conference in Meringa. Froggatt was appalled; he saw the introduction and release of Cane Toads as the height of idiocy. Blocked by politics from publishing a letter to that effect in the Sydney newspapers, Froggatt sent his letter to a Melbourne newspaper instead. He pressured the Australian quarantine authorities to ban any further release of Cane Toads.

Initially, Froggatt's crusade succeeded. The quarantine authorities banned further releases, noting that "the toads might become an even greater pest than the cane beetles they were to destroy." Truer words were never spoken, but truth rarely plays a major role in political battles. Froggatt was a voice in the wilderness. The pro-toad Queenslanders had support not only from the Hawaiian heavyweight (Pemberton), but also from the cane growers, who were keen to flex their political muscle. A government minister, the delightfully named Frank Bulcock, fumed that "the toads are the best economic contribution to the sugar industry for many years, and I shall take up the matter with the Federal Government at once." The national scientific authority CSIR (the forerunner of today's CSIRO, or Commonwealth Scientific and Industrial Research Organisation) backed the cane growers as well. CSIR wanted to import toads themselves: a European species, to control insects in cooler parts of the country. A ban on Cane Toads would derail their plans. The one-sided battle was over by the end of 1935. Prime Minister Joseph Lyons overruled his quarantine officers, and the ban on Cane Toad release was lifted.

But the enmities still rankled. Pemberton wrote of his old friend:

> I am surprised that a man of Mr Froggatt's standing as a naturalist could entertain such radical and biologically impossible apprehensions respecting so beneficial and innocuous a creature as Bufo marinus. It is further surprising to me that he should write in such a vein, since I discussed the whole matter with him at great length on several occasions. I told him of the objections raised in Honolulu against it by a few neurotic old women and an occasional chronic pessimist, who cared nothing for the benefits accruing to the

general community, but only conjured in their own selfish minds dire results to their nervous systems should one of the "loathsome creatures" cross their path in the rose or vegetable garden when in pursuit of a cockroach, centipede, scorpion, or pestiferous beetle.

And in a thinly veiled attack on Froggatt, Frank Bulcock characterized critics of toad importation as "fortified by a colossal ignorance." A lovely phrase, but history tells us that the Champions of the Toad were wrong.

The passions are evident in a talk that Reg Mungomery gave at an international sugarcane meeting. Speaking to a supportive audience, Reg said:

> Much has been printed in the newspapers during the last six months concerning the activities of toads, and in particular an undue amount of publicity has been given to details concerning their poisonous attributes. Certain newspaper correspondents have been over-ready to express their ideas on the habits of toads, their poisonous properties, and the supposed evil effects following the introduction of this toad into Australia, and their stories have been published. Such stories are reminiscent of ancient and mediaeval times when the toad was considered a deadly enemy. It appears to have always been the subject of wild surmise, and as one writer aptly says, "With no other creature has invention been more free".... Although false reports have been denied officially... several people who have never read these denials... still consider the toad a dangerous importation.... I have had reports of toads being killed by small boys and others.... I enquired if we might threaten the offenders with prosecution under the Sanctuary Act, but learned that unless the toads were killed by gunfire no action could be taken.... In order to protect the toads from similar unwarranted destruction, the only course now open is to have the toads protected under Government proclamation, and if the delegates to this conference see fit they might pass a resolution asking that these toads be put on the list of protected fauna under the "Animals Protection Act of 1925" or the "Fauna Protection Act of 1937."

Two things are clear from this diatribe: Reg's resentment of Walter Froggatt (the "newspaper correspondent") and his concern about threats to his beloved toads. It beggars belief that an authoritative scientist would petition the government to declare Cane Toads a protected species. And in the grasp of that desperation, Reg came perilously close to a bare-faced lie. He knew full well that toads were toxic, but he ridiculed the suggestion ("The idea ... that a person has felt ill after handling a toad is purely a figment of his rather fertile imagination").

I agree with at least one point from Reg's monumental dummy-spit: toads do attract nonsense. Misinformation about Cane Toads has been a staple fare of the media for decades, has sold a lot of newspapers, and has increased ratings for many TV shows. And things have not improved since 1935.

Overall, then, the Cane Toad's introduction to Australia was not just a poor decision by a couple of cowboy scientists. Seduced by a powerful industry lobby, the political leaders of the country condoned environmental vandalism.

The Cane Toad's public popularity soon dissipated: it was welcome in the cane fields, but not around the towns. The last reference I can find to applaud the toad ("An American doing good work among the cane pests in North Queensland") was written in 1943. Meanwhile, disquiet was growing farther south. The beekeepers of New South Wales demanded that the government stop the toads from spreading past the border. Toads often gather around beehives to pick up a free meal. A newspaper article of 1946 related that "when fully grown, these toads were capable of reaching up to fully a foot, and would weigh anything from five to seven pounds with a tongue six inches long that swept bees off hive entrances in alarming fashion." Alarming, indeed, had it been true.

The media began to express revulsion at the toad's appearance as well as its size. An article titled "Giant Toad Comes Marching South," in the *Sydney Morning Herald* of August 2, 1947, began with this stirring paragraph:

> Everything about it—its very name—is vile. Its huge belly is always distended. The angry glitter in its eye is revolting. It has legs but they do not raise it from the mire. It has eyes but they do not welcome light—its bane. Its habitation is filthy. Its habits disgusting. Its color dingy. Its breath foul.

They really knew how to write newspaper articles in those days, although I doubt that the author had ever smelled a toad's breath. The not-so-subtle motive for these diatribes was interstate politics: the opportunity for New South Wales to blame Queensland for introducing the toad.

Regardless, the horse had bolted. By the time that Walter Froggatt questioned the consequences of bringing Cane Toads to Australia, the alien amphibians were already breeding in the Queensland cane fields. Sporadic attempts to slaughter toads provided a sad foretaste of the futile toad-killing extravaganzas that would grip tropical Australia decades later. It was a waste of time. Cane Toads were in Australia, and they were here to stay.

Over the next few decades, the toads spread slowly down the Queensland coast, and rapidly westward toward the Northern Territory. Defying all predictions, they crossed the harsh Barkley Tablelands and reached the border of the Northern Territory in 1983. Moving faster and faster, they spread through coastal regions of the Northern Territory and across the border into Western Australia in 2009. By 2016, Cane Toads occurred over more than 1 million square kilometers (62,000 square miles) of Australia—more than an eighth of the entire continent. And, ominously, they're still spreading.

Although the Cane Toad's conquest of Australia now features in textbooks of ecology and evolutionary biology, the toads haven't read those books. If they had, they would have realized the impossibility of what they have achieved. Even a Cane Toad would have given up when it understood the challenge lying before it.

Under Charles Darwin's theory of evolution, we expect all organisms to be intricately adapted to the places in which they live. The reason is simple. Put yourself in the animal's place—any type of animal will do. Imagine that you have inherited a new characteristic (caused by a random mutation, a change to your genetic constitution) that makes you better suited to the challenges that you encounter. It doesn't matter if those challenges come from climate or from other species; all that matters is that this new change helps you survive. It's just pure chance that you happened to be the lucky beneficiary of that genetic novelty; it didn't arise *because* it would help you. Instead, it arose as a happy accident—an error when your parents' genes were being copied into the germ cells that formed you as an early embryo. Because the characteristic is genetically based, you will pass it on to your offspring. They will also do well in life (because that genetic change increases their success, as it did yours), so that the number of individuals carrying that favorable mutation will increase. You'll have many healthy grandchildren.

In contrast, imagine the opposite situation, where you inherit genes for a trait that makes you *less* well suited to your environment. Bad news. Those genes won't leave copies of themselves, because their bearers die before they can reproduce—and as a result the new, unfavorable mutation soon disappears. No grandkids. That's what evolutionary biologists mean when they talk about "survival of the fittest" and "natural selection." It's a simple process that adapts organisms to local environments. Traits that help an animal survive become more common; traits that are a disadvantage become less common.

It's clear to everyone (even creationists) that animals exhibit complicated, elegant adaptations to deal with the conditions they encounter. Every wildlife documentary has examples of animals overcoming such challenges. But there's another side to the coin; what works in one place isn't likely to work in another. If you evolve in Barbados, you won't thrive in Alaska (and vice versa). A tropical species needs to function at high temperatures, whereas an Arctic species has to deal with extreme cold. If you can do one of those things, you probably can't do the other.

After evolving to thrive in the moist rainforests of Latin America, surely a Cane Toad would shrivel up and die in the Australian outback? As the descendants of those first Cane Toads marched westward, why didn't they stop and turn around when they hit the arid inland plains? A sensible amphibian would have headed back to the well-watered coast, where conditions resemble their South American homeland. But that didn't happen. The toads continued their westward march, crossing the arid landscapes of the Gulf of Carpentaria during brief periods when the ground was moistened by heavy rain.

The world encountered by those first Amero-amphibian invaders of the Northern Territory was hospitable during part of the year, but unfriendly the rest of the time. The toad's invasion across Queensland took them from the coastal "wet tropics" (where it rains year-round, to the dismay of tourists) through to the "wet–dry tropics" of coastal Northern Territory and Western Australia. In this region, there is a lot of rain—heavy rain—for a few months each year; Darwin is green and lush from January to March. But you won't need an umbrella if you visit in June. The skies are cloudless. The soggy conditions of the wet season give way to an increasingly parched landscape, the Toad Heaven of February (wet soil, and bugs aplenty) turning into Toad Hell (dry, hot, and barely a bug to be seen) by May or June.

To understand the challenge faced by those original toads, take a walk around the east-coast rainforests of Cairns, then do the same 3,000 kilometers (1,800 miles) to the west, in the Kimberley. In the wet tropics of north Queensland, where Reg Mungomery liberated his young toads, conditions are comfortable. In the shade of the forest, a Cane Toad can sit out in the open during the day. Pleasantly warm, always moist—a place where a South American amphibian can kick back and relax. Things are very different in the harsh, sun-baked, rocky hills on the western edge of the toads' great adventure around Kununurra, or on the stony plains around Longreach in central Queensland. If it sat out in the open by day, even a tough-as-nails Cane Toad would be dead within an hour. The lush green rainforest of Cairns is amphibian nirvana, but the outcrops of the Kimberley are hell for a small creature

Despite their evolutionary origins in moist habitats on the fringes of the Amazon rainforest, Cane Toads have penetrated into highly arid regions of Australia. I took this photo near Longreach in central Queensland.

with moist skin. Frankly, they're a challenge even for a relatively large professor with dry skin.

The highest average monthly temperature in Cairns is 31°C (88°F), whereas Kununurra hits a brain-numbing 39°C (102°F). Two weeks without rain has the residents of Cairns complaining about drought, whereas Kununurra is dry from April to November. Very few frog species are found all the way from Kununurra across to Cairns. It takes a special amphibian indeed to survive in such different places. And remember, native frogs have had millions of years to evolve solutions to those problems. Cane Toads have done it in the blink of Charles Darwin's eye.

Native frogs cope with months of hot and dry conditions by waiting it out—sitting on their little backsides until it rains. The amphibians of tropical Australia are the children of El Niño, the climatic oscillation that brings torrential rain in some years and drought in others. The native amphibians live by a simple rule: snooze when it's dry and party when it's wet. They spend the long dry season deep within burrows, shutting down any physiological system that costs energy. Eons of evolution have fashioned an animal that is active for only part of the year, just as reptiles are in cold climates.

Unlike native amphibians, Cane Toads don't slow down when the rain stops. The wider landscape is too parched for them to disperse, but they remain active. While the native amphibians are in a near coma, in suspended animation until the monsoons return, Cane Toads shuttle backward and forward from their daytime shelters to the dripping tap, and then to the night-light that attracts juicy bugs. The ancestors of Cane Toads evolved in habitats that are wet most of the time, but with unpredictable—and usually brief—periods of dry weather. Even in their native range, Cane Toads sometimes have to spend a few weeks in a moist refuge site until the next shower. As a result, Cane Toads are flexible. They hunker down in dry weather, but they are ready to move out as soon as things improve.

If Cane Toads had arrived in Darwin in the nineteenth century, the tactic of continuing to move about and feed during the dry season would have been suicidal. But humans have changed the landscape to make it toad friendly. We've hung out a large "Welcome" sign across tropical Australia, turning an inhospitable landscape into one that suits our warty invader down to the ground. When they arrived in Darwin in the twenty-first century, Cane Toads felt right at home.

The biggest change was in access to water. Ponds were rare in the natural landscape, but today they are everywhere. Not only farm dams and stock-watering troughs, but also many other damp places where a dried-out toad can find salvation. Every house has a cool, moist space beneath the floor—a delightful motel room for a wandering toad. And at night that recently watered lawn, or the dripping hose beneath the air conditioner, is a magnet for toads. Even the dog's watering bowl will enable a toad to survive dry weather.

And we provide toads with food as well. The bright lights around our houses are beacons that attract bugs from far away. It's an amphibian's smorgasbord, with dazed bugs cannoning off the light and falling to the ground. And the paddocks of cattle properties are littered with moist cowpats that attract dung beetles. What more could an immigrant amphibian ask for?

Our generosity to the toad extends to their travel plans. Like us, toads can move more quickly and easily along a road than through the bushland. And if that's not enough, how about a free ride? Toads love to hide inside objects on the ground, as I've often discovered when pulling on my boots in the morning. Cane Toads frequently stow away on trucks, inside loads of building and landscaping materials. Other toads are shipped around on purpose; the first twenty Cane Toad colonists of the Northern Territory were imported for school students to dissect in 1974. The amphibians escaped from the teacher's house but died in the harsh, dry-season conditions. A few weeks earlier, a bunch of toads had escaped in Perth under similar circumstances. Ever since Reg Mungomery provided a free trans-Pacific trip, Cane Toads have been catching a ride with humans.

And emulating Reg's pampered pets, hitchhiking Cane Toads even embark on international adventures. One New Zealander was horrified to find a Cane Toad inside his boot when he returned home from a holiday in Queensland. Sydney locals were shocked to see toads emerge from a shipping container brought from Fiji. The record for long-distance travel, though, goes to two Cane Toads that stowed away in a Ford Mustang in Florida. The car was shipped by sea to England, a distance of 5,000 kilometers (3,000 miles), by which time the toads were emaciated but otherwise healthy. So a quick road trip from Brisbane to Sydney is just a comfortable outing for these rugged amphibians.

Most stowaways come to a quick and lonely end, but some settle into their new homes. The southward spread of toads along the coast of New South Wales has been due mostly to people moving toads around, rather than the animals dispersing under their own steam. Whether those translocations were due to happenstance, or to a misguided hippie trying to control the bugs in her marijuana plantation, may never be known.

It wasn't enough to subsidize the Cane Toad's needs for water, food, and transport. We've also enriched their sex lives. In the Australian tropics, any depression made by bulldozers during the dry season fills up with water when the monsoonal rains arrive, creating a perfect sex palace for toads. Native amphibians prefer ponds with deep water, shade, and vegetation rather than these open, shallow scrapes. But Cane Toads like to breed in the open.

And, last but not least, humans have decimated the toad's main competitors: native amphibians. Sometimes that's been done by bulldozers and water pollution, and sometimes by feral cats and pigs. Humans have also spread diseases that are lethal to frogs. Because the tadpoles of native amphibians compete with Cane Toad tadpoles, removing those native tadpoles leaves more food for the toads.

The assistance to Cane Toads extends to environmental legislation. For example, when Cane Toads bred in an industrial site in suburban Sydney, the local council tried to eradicate the intruders. But it wasn't easy, because Cane

Toads aren't listed as a noxious species under New South Wales law—so when the factory owners were unhelpful, the council rangers couldn't access the pond where the toads had spawned. Nor could the council prevent an ideal toad-breeding pool being constructed a year later on the property next door. In the end, the red tape helped nobody but the toads.

Thus, we've changed tropical Australia to make it the ideal "home away from home" for American Cane Toads. We've provided water, food, places to breed, and roads to speed their invasion, and we've eliminated their competitors. Despite rhetoric about "stopping the toad," *Homo sapiens* has been the Cane Toad's greatest ally.

Sydney is supposed to be too cold for Cane Toads to breed, but they did it. It makes me wonder how far Cane Toads will eventually spread in Australia. Don't underestimate these expert invaders. Because toads are more than willing to hitch a ride, every part of Australia will receive an occasional warty visitor. That's not a major problem if the toads arrive in small numbers in the city suburbs (the destination for most trucks), where large native predators are already scarce. That solitary, wandering toad may kill a fox terrier, but not a marsupial. The more important question is this: Where will toads be able to set up populations that will remain viable in the long term?

Scientists have tried to predict the eventual limits of Cane Toad distribution in several ways. The first attempts were based on "climate matching": which parts of Australia offer the same kinds of climatic conditions that are occupied by Cane Toads in their native range? Unfortunately, that method depends on two assumptions. First, that the Cane Toad's distribution in its native range is limited only by climate. If competition with other species of toads excludes Cane Toads from some areas of South America, for example, "climate matching" will underestimate toad spread in Australia. The second assumption is that Cane Toads haven't evolved during their Australian odyssey; if they can adapt to dry conditions, for example, then their distribution in Latin America is no longer relevant. In fact, toads have already colonized parts of western Queensland that are hotter and drier than their native range. So the "climate matching" approach to predicting Cane Toad distributions in Australia is looking flimsy.

The next generation of modeling efforts solved that problem by broadening the definition of "suitable climates" to include areas that the toads already occupy in Australia. That method prophesied a wider distribution for the toad, but there was still a problem: it's impossible to summarize "climate" in just a few measures. For example, based on average annual temperature and rainfall, we'd expect the suburbs of Perth to be fine for Cane Toads. But most of that rainfall occurs in winter, when it's so cold that any sensible Cane Toad would be hiding deep in a burrow. As a result, Perth doesn't provide the hot, wet conditions that toads need. Or to take the opposite case: climatic data for inland Queensland would tell us that rainfall is far too low for Cane Toads. No need to worry. But we'd be wrong: those inland rivers carry water from downpours that fell hundreds of kilometers upstream, creating oases in otherwise inhospitable downstream areas. Cane Toads don't care where the water comes from.

More recently still, Mike Kearney of Melbourne University has developed another way to predict toad distributions. This approach builds a detailed picture of toad biology to estimate where a Cane Toad can make a living. Mike's mathematical models identify how many hot, wet nights a toad needs each year to obtain enough food to grow and reproduce; and whether water for breeding is available, and for how long. His model then uses climatic data to calculate whether or not a particular area is suitable for Cane Toads. It's more complicated than the earlier methods, but it gets closer to the truth.

Can the citizens of Perth sleep soundly, knowing that their city is out of the toads' potential range? Not necessarily. Cities are toad magnets. First, they provide water in the form of ponds, dams, and dripping taps. Second, they provide edible bugs that flourish in well-fertilized, well-watered gardens. Third, cities are safe havens because the toad's potential competitors and predators have been eradicated. And fourth, cities are warmer than the surrounding countryside because of the "urban heat island" effect. Electricity-devouring machines produce local hot spots. Even if the surrounding climate is too cold, too dry, or bug free, humans create a Cane Toad Paradise within suburbia.

Nonetheless, humans don't deserve all the credit for the success of the Great Toad Invasion. The toads themselves have done a fine job. They arrived in Australia with a talent for flexibility, and they built on that talent by evolving to overcome every new challenge that appeared. Because a Cane Toad can produce 40,000 eggs per clutch, there is a lot of variation among offspring for natural selection to act upon—and adaptive changes can thus happen quickly. And the harsh conditions of Australia impose strong pressures; very few of those 40,000 eggs survive to become adult toads. As a result, any favorable mutation that arises soon becomes more common, adapting the Cane Toad to life in Australia.

The Cane Toad's ability to deal with new challenges is most obvious at the fringe of the Australian desert. For an arid-zone toad, life revolves around the farm ponds or troughs that provide water for thirsty cattle. In well-watered regions of coastal Queensland, a toad can spend the day almost anywhere—even in a clump of grass. But a toad in the desert needs to be back in the pond by sun-up or it's doomed, like the trolls of medieval legend. There is nowhere else in that dry landscape where it can survive. Toads in the desert live on the edge, with no safety margin if things go wrong.

Faced with the constant danger of desiccation, Cane Toads have developed some novel counterstrategies. For example, stressed animals (of all species) respond by producing a hormone called corticosterone—and as we'd expect, the desert toads are full of it. But, intriguingly, those high stress-hormone levels also cause the toads to retain water—thereby helping them stay alive in this inhospitable country. And out there on the fringes of the Tanami Desert, a toad will do whatever it takes. If it's inside a soil crack that is drying out too fast, the toad will climb out into the harsh midday sunlight to sprint toward the nearby dam. Everywhere else they live, adult Cane Toads are nocturnal—but a toad that is running out of water knows that it has to break the rules and move around by day. In our lab trials, Cane Toads from relatively benign environments stop moving around when they begin dehydrating; they just sit in the corner and wait for the kinder (moister) conditions that are likely to be on their way. In contrast, desert-dwelling Cane Toads keep moving even after they have lost more than 20 percent of their body water stores. If you stop before you reach the pond, you die. The desert toads also take up water more rapidly through their skin, a useful trick to hydrate in a hurry.

This ability to adjust to local conditions is a problem for crystal-ball-gazing exercises to predict the limits of Cane Toad distributions, with respect to temperature as well as moisture. For example, if Cane Toads don't occupy any area that falls below 10°C (50°F) in their native range, will the same be true in Australia? No, it won't: toads in southern Australia are already colonizing places that we thought were too frigid for them. They tolerate these conditions by adjusting rapidly: a few hours of exposure to cold improves a Cane Toad's ability to handle low temperatures. We don't know how they do it, but this mysterious mechanism allows toads to climb high mountains in northeastern New South Wales. Every spring, toads head up the slopes. Many of them die up there next winter, but there are always more toads arriving to replace them. That's bad news for predators in the mountains. Even if your home is too cold for Cane Toads in winter, toxic packages of hopping death may arrive on your doorstep in spring.

Where does this leave us in predicting the eventual range of toads in Australia? If you live in Hobart or Canberra, don't worry. I can't imagine a tropical amphibian surviving in such a cold place. Indeed, I'm surprised that human beings can live there. But if you reside in Perth or Adelaide, don't be too astonished if one day, a Cane Toad arrives at that swamp down the road.

How have Cane Toads affected ecosystems in Australia? That's been a popular topic of conversation in committee meetings and hotel bars ever since Cane Toads began to spread out of the Queensland cane fields. In the early days, the debate was simple. On one side, the Sugar Mafiosi claimed that toads were not a problem. As Reg Mungomery confidently asserted,

> To others who ... suggest the possibility that the toad will, in turn, itself become a pest, we can point to the fact that nearly 100 years have elapsed since it was first introduced into Barbados, and there it has no black marks against its character. Experience with it in other West Indian islands, and in Hawaii, certainly points to the fact that no serious harm is likely to eventuate through its introduction in Queensland.

On the other side of the debate, the lone voice of Walter Froggatt predicted that

> All our ground fauna will become their prey, and all our curious, mostly harmless, and often useful ground insects, in forest and field, will vanish. The eggs and nestlings of all our ground nesting birds will be snapped up by these night-hunting marauders. All our frogs and lizards, most valuable insectivorous creatures, will be in danger of their lives.... [T]his great toad, immune from enemies, omnivorous in its habits, and breeding all the year round, may become as great a pest as the rabbit or [Prickly Pear] cactus.

Although he worried that the toad would devastate the Australian fauna by eating it, Froggatt didn't recognize an even bigger problem: a native predator that tries to eat a Cane Toad is likely to be fatally poisoned. The toad's toxins disrupt fundamental cellular processes, interfering with the mechanism that pumps sodium and potassium ions across the cell wall. Most Australian reptiles suffer a heart attack as soon as they swallow even a few drops of toad poison. The importance of lethal toxic ingestion was finally recognized in 1975, when two biologists at the Queensland Museum, Jeanette Covacevich and Mike Archer, connected the dots. Mike was stung into action when his pet quoll (a cat-sized marsupial) died after eating a Cane Toad in Mike's suburban backyard in Brisbane. When I asked Mike about it, a quarter of a century later, his distress at the quoll's death was still evident.

Why did it take so long to recognize the danger posed by toad poisons? First, because these impacts hadn't occurred when Cane Toads were brought to places without native predators (like Hawaii), or where coexistence with other toad species had adapted the local predators to toad poisons (like Puerto Rico). But Australia, unique in its lack of native toads, was a different proposition. The second reason was that most of the toad's victims in Queensland were reptiles, unpopular with the general public. The impact of the invaders' parotoid secretions fell on "useless" animals.

As a result, toad toxicity stayed below the political radar. In July 1947, Norman Laird wrote (in a Tasmanian newspaper) that "snakes ... are poisoned when their mouth parts come in contact with the toad's skin." In 1952, a Melbourne newspaper *(The Argus)* reported that "Giant toads are attacking the deadly Taipan snakes in the Mackay sugarcane fields. The honors so far are even. Dead Taipans and Death Adders have been found in the fields with dead toads nearby. . . . Cases are on record of dogs and cats dying after attacking the toads." The article continued: "A Mackay naturalist, Mr. J.H. Williams, is anxious for eye-witness account of fights between the toads and the snakes to find out which is the aggressor." Incredibly, naturalists thought that toads might be attacking snakes, rather than (as was undoubtedly the case) vice versa. Similarly, male toads found clutching female frogs were interpreted as would-be murderers, rather than as visitors trying to become more intimately acquainted with the locals.

Early media reports treated Cane Toads as topics of quirky interest, with no mention of their negative impacts on native animals. Since those times, though, the message has changed. An avalanche of publicity has convinced Australians that Cane Toads are exterminating our wildlife, creating ecological Armageddon. Scientists contributed to that dismal forecast, in an attempt to wake people out of their complacency. For example, in 2002 the doyen of Australian frog biology, Mike Tyler, was quoted (in the *Guardian* newspaper) as saying, "You can now forget about Kakadu. Kakadu is lost. Cane Toads are going to be more prolific here than elsewhere, and they're going to get bigger. It will become the most dominant form of life in a little bit of Australia that we thought was pristine." Mike's bleak outlook encapsulates the collective despair of Australian ecologists (myself included) at that time.

But it was guesswork at best. We didn't know how badly Cane Toads would affect Australian ecosystems, because all we had were anecdotes. Dead snakes in the cane fields or on the floodplain, with toads in their mouths. Impressions that native species became less abundant after toads arrived. But no matter how often you tell me the story about Aunt Mary finding a dead goanna beside the chicken-shed soon after Uncle Ken thumped the first-arriving Cane Toad into pulp, it doesn't qualify as scientific evidence of toad impact. And that's a hugely important knowledge gap.

Measuring the ecological effects of a biological invasion isn't easy. We can't just go out and count animals before and after, and then attribute any change to the invader's arrival. Wildlife populations fluctuate in response to many factors. For example, the amount of rain during the wet season in tropical Australia influences feeding opportunities for native fishes, frogs, lizards, crocodiles, snakes, wallabies, and so forth. As a result, a sudden decline in the numbers of one species—say, Frillneck Lizards—is as likely to be due to poor rainfall as it is to toad arrival. Teasing those factors apart requires information, not anecdotes.

Because nobody had conducted robust field studies on the toad's impact, we had to rely on circumstantial evidence. And things looked dire. In 2003, one of the first scientific papers I wrote on Cane Toads was based on laboratory trials in which we gave tiny doses of toad poisons to snakes. Most snake species were highly sensitive. We predicted that "Cane Toads threaten populations of approximately *30%* of terrestrial Australian snake species." Fortunately, we were wrong. The next chapter tells the story of how we discovered the reality of toad impact.

THREE

Arrival of Cane Toads at Fogg Dam

THE BIRTH OF TEAM BUFO

> Scientists' minds may jump around like amorous toads, but they do seem to accept such behavior in one another.
>
> CALEB CARR, *The Alienist*

Until Cane Toads hijacked my career, my research was focused on three questions: What do animals (especially snakes) do in the wild, why do they do it, and how can that knowledge help us conserve them? The phrase "why do they do it" tells you that my interests lie at the intersection between ecology and evolutionary biology—both broad fields in their own right, encompassing many approaches that have never appealed to me. For example, I've ignored "community ecology"—the field that asks how groups of species coexist with each other. I have friends who are passionate about that question, but it's just too messy and vague for me. Likewise, classical evolutionary biologists often focus on discovering how species are related to each other, and how to tell them apart. But cramming biological variation into pigeonholes doesn't light my fire either.

Instead, I have spent most of my career trying to understand *why* reptiles have evolved as they have. Why are males larger than females in some snake species, but smaller in others? Why do some lizards give birth to fully formed offspring, whereas others lay eggs? Why is a turtle's sex sometimes determined by its chromosomes, and sometimes by incubation temperature? The answers to those questions depend on how an individual's characteristics influence its chances of living and breeding—that is, natural selection. And so I explored how snakes function in the wild. I spent four decades investigating the evolutionary advantages and disadvantages that have shaped the bodies, behaviors, and ecologies of snakes (and, occasionally, other reptiles). It was enormous fun.

That background directed my Cane Toad research in unorthodox directions. Most research into "invasion biology" has been purely ecological.

Believing that evolution was too slow a process to worry about, scientists largely ignored the role of evolutionary processes in shaping the course of an invasion. But soon after Team Bufo took on the Cane Toad, we realized that both toads and their victims are evolving rapidly in response to the new challenges they face. My background in evolutionary biology rather than "classical ecology" helped us zero in on some fascinating features of the Cane Toad invasion.

But my leap into Cane Toad research wasn't easy. I had no desire to switch from long, slithery creatures to small, round, hopping objects. Snakes are elegant and mysterious, whereas frogs are merely cute. For the first few decades of my career, my only research involvement with frogs had been when I found them inside the stomachs of preserved snakes. Being partly digested doesn't improve anyone's attractiveness, so those little half-eaten corpses didn't enthuse me to go out and study frogs. I thought of amphibians as snake fodder—but that all changed in 2001, when Cane Toads entered my research world.

My first foray into toad biology arose when a young man named Ben Phillips asked me to supervise his Ph.D. studies and I agreed. Ben has been a core member of Team Bufo ever since. He is part hippie, part field biologist, part mathematical modeler, with a happy-go-lucky attitude and an engaging giggle that distracts you from his formidable intellect. Conversations with this ludicrously intelligent polymath identified several potential projects, but we ultimately agreed that the most exciting topic for his Ph.D. would be to explore the impact of Cane Toads on native snakes. Many people were forecasting ecological catastrophe, but the evidence was meager. Ben approached his project in novel ways, including buying an old van and driving across Australia to catch snakes and test their tolerance to toad poisons.

After two years' work, Ben concluded that Australian snakes were in dire straits. His laboratory tests showed that most Aussie snakes (especially the venomous ones of Family Elapidae) were very sensitive to bufotoxins; they would die if they ate even a small toad. So we wrote a paper in 2003, foretelling doom and gloom. Australia's snakes were in trouble.

I enjoyed working with Ben, but it was a sideline to my main research. Ever since the mid-1980s, I had focused on the ecology of snakes on the Adelaide River floodplain, 60 kilometers (40 miles) from the city of Darwin, in the Northern Territory. To get there, you drive east out of Darwin for an hour, through the town of Humpty Doo, then turn north toward a dam—Fogg Dam—that was built in 1956 to provide year-round water for commercial rice growing.

This bold experiment in tropical rice growing was big news when I was a schoolkid in the 1960s. A consortium of Australian and American investors decided to improve on the laborious Asian methods of growing rice via manual labor; instead, new-fangled machines could sow and harvest the crop. At school, we were told that mechanized agriculture would transform tropical Australia into a massive food-producing center. It never happened. The history of agriculture in the Northern Territory is a litany of failed dreams. I've seen farmers try bananas, mangoes, melons, Water Buffaloes, and many other products—but it's a long way to the southern markets, it's hard to find a labor force, and the viruses, insects, rats, and birds eat anything that grows.

The rice-growing enterprise failed after a few years, but its legacy fashioned my research career. A small village with the prosaic name of Middle Point was built only a kilometer (half a mile) from Fogg Dam, to house the government biologists whose research would help the rice growers combat the geese and rats waiting to consume their crops. Like the farmers, the scientific team were soon gone—but they left behind a research facility nestled beside the greatest snake pit in Australia.

Fogg Dam was created by constructing an earthen wall across the floodplain. Built to irrigate the rice crop in dry weather, the dam was never used for agriculture. Instead, it became a magnet for wildlife. If you ignore the heat and humidity, and cover yourself with insect repellent, Fogg Dam is paradise for a biologist. And that's especially true if, like me, you harbor a deep-seated, illogical, but absolutely unshakeable passion for snakes. Fogg Dam is Snake Heaven. For most of my career, from 1985 to the present day, it has been my second home. I spend several weeks there each year, mostly delving into the private lives of snakes.

The studies went well but were frustrating. Zipping up to the tropics for a few weeks at a time gave me staccato glimpses of the ecosystem, but I would never truly understand it without full-time data. Finally, in 1989, I obtained a major grant from the Australian Research Council, enough funding to employ a postdoctoral researcher for the Fogg Dam project.

Hiring somebody for the job was daunting: not many people are tough enough and enthusiastic enough to conduct fieldwork year-round in tropical Australia. But I was lucky—I found two ideal candidates. A young scientist named Thomas Madsen took over the Water Python project in 1989, transforming the project from "pith-helmet biology" to a full-time study. Thomas has since taken another job, but we still collaborate on the python research. It took nine more years for me to rake up the cash for another researcher's

A Freshwater Crocodile from nearby Fogg Dam strolls past the Middle Point research station on its way to resting for a few hours in the shade of our carport. Photo by Greg Clarke.

salary, but Greg Brown joined the Fogg Dam team in 1998 to focus on the smaller snake species.

I kept coming up from Sydney whenever I could, for a few weeks at a time, juggling the demands of undergraduate teaching and a young family. In 1994 my family came with me, and my two sons spent a few months enrolled at the local school. Middle Point became my home away from home, and Fogg Dam occupied an increasingly central place in my heart.

Strangely for tropical biologists, both Thomas and Greg hail from colder lands. Thomas is a Swede, and Greg is a Canadian. Most people find the Fogg Dam heat and humidity hard to bear, but Greg is the exception. He loves it. He spent the first twenty-five years of his life dreading the Canadian winter and feeling chilly even in summer. He claims to have written his Ph.D. dissertation while lying in a bathtub full of hot water. If the temperature at Middle Point drops below 32°C (90°F), Greg dons a heavy jacket.

The project that brought Greg to Fogg Dam was focused on snakes—nonvenomous ones, because Greg avoids anything that might kill him (a refreshingly rare attitude among snake enthusiasts). The centerpiece of Greg's snake research (which continues, almost twenty years later) is a "mark–recapture" study. It's simple stuff. You visit an area repeatedly, catch every animal you can, give it an individual mark so you can recognize it again, and then let it go where you caught it (after measuring its size, recording its sex, and so forth). And, with luck, you catch it again—perhaps many times. It's straightforward and low-tech but provides fundamental information. For example, you find out how many animals live in the area, how quickly they grow, how far they move, how long they live, and how often they reproduce. And in an area like Fogg Dam, where rainfall patterns change from one year to the next, you find out how the population responds to that environmental variation.

If you stroll along the wall of Fogg Dam at night (carefully, given the Saltwater Crocodiles that might be lying in ambush), you will encounter a thousand species of insects but only three species of nonvenomous serpents. Soon after he arrived, Greg began walking the dam wall in the evening, catching snakes. He brought them back to the research station, measured and marked them the next day, then released them the following night. Few people throw themselves into their work as enthusiastically as Greg did. He decided that catching snakes a few nights a week wasn't enough; instead, he did it *every* night. When I came up to Middle Point in 1999, a year after Greg had started, he hadn't skipped a single night on the dam wall in the last several months. And the same thing happened the next year, and the one after. It's still happening. This means three things. First, we have a vastly more detailed picture of wildlife activity on the wall of Fogg Dam than we have for almost any other place on Earth. Second, Greg is either highly dedicated or seriously disturbed. And third, his wife Cath deserves sainthood.

Greg doesn't stand out in a crowd: medium height, medium build, with curly hair that was once brown and is now turning gray (with a brief intervening period when he tried life as a blonde). Greg's tastes are eclectic. His clothing is purchased from the second-hand sale rack at the local charity outlet, apart from a few tattered garments dating to his student days. But his genial smile hides a photographic memory about abstruse issues. Without Greg's store of information, I wouldn't know that Alexander the Great's body was preserved in a vat of honey, or the four minerals from which Captain Kirk fashioned a bazooka to defeat the Gorn lizard-monster in episode 18 of *Star Trek*. Interested in the history of Spartan warriors? So is Greg; he avidly reads anything written about them. Or perhaps Johnny Winter's music, or a Monty Python movie? I bet that Greg recalls those chords, plots, and characters better than you do.

Like many Canadians, Greg is unfailingly polite. In a rough-and-tumble Northern Territory culture where even the nuns and schoolgirls mutter

The founding members of Team Bufo, left to right: Ben Phillips, me, Michael Crossland, and Greg Brown. Photo by Terri Shine.

obscenities, Greg rarely says anything more steamy than "holy moly." He's a recluse, and his idea of nirvana is a total lack of other human beings (except Cath), but enriched by the presence of Water Buffaloes, turtles, and toads. Greg's most unusual traits are that (1) if he doesn't have anything to say, he doesn't say anything, and (2) he thinks about a question before he answers it. Most of us fill those silences with small talk, but not Greg. It's fun to watch newly arrived students ask Greg a question, then see them begin to squirm as he stares off into the middle distance for a few minutes, framing a well-considered answer.

The research station at Middle Point was ground zero for our Cane Toad research, so it's worth describing what the place looks like and how we conduct our studies there. Middle Point gets its name from its location—it lies on a thin finger of high, dry land projecting into the floodplain of the Adelaide River. The village consists of five houses (one of them derelict) and a single block containing four apartments, plus a central building that contains our offices, laboratory, and workshop.

Although Middle Point is an hour's drive from anything you could call civilization, it's not set among pristine wilderness. Fogg Dam is surrounded by a conservation reserve, and it's only half a mile away from the village—but the rest of the landscape has been cleared of trees and is used for Water Buffalo farms, a crocodile farm, a fish (Barramundi) farm, and mango orchards.

To a newcomer's eyes, the scraggly eucalypt forest of the Top End is likely to be disappointing. It's a monotonous vista, with only an occasional bright-green Ironwood Tree or Cycad to break the gray-green dominance of the Stringybark and Woolly-butt trees. A roadside sign, adorned with a cartoon Frillneck Lizard, proclaims "We like our lizards frilled not grilled." The sign is difficult to read, because it's been scorched by a fire. Shade is rare, and wildlife is even rarer—except for ants. An area the size of a football field contains more than a hundred different species of ants, a number that makes vertebrate biodiversity look feeble. Unless you spook an Agile Wallaby at rest under a shady Ironwood, though, anything bigger than an ant is hard to find.

If you want to capture the flavor of life at Middle Point, hop into a sauna or steam room and liberate a few thousand biting insects, then dribble some insect repellent and sunscreen into your eyes. It's always hot, although the nights are mercifully cool (Greg would say "cold") midyear. November is a killer. It's the buildup season, when humidity and cloud cover increase but it doesn't rain. So it's always very hot, and always very damp, and you watch the mold spread across the ceiling day by day. If you don't drink vast amounts of water, you'll develop a killer headache.

The mosquitoes come out in force at sunset, but a liberal dose of repellent keeps them at bay. Having been cooped up all day in an air-conditioned office, I often stroll over in the evening to greet Greg and Cath's pet Water Buffaloes. Mary, Joe, and Bobo were abandoned by their mothers and have been pampered from tender calfhood. Their owners give them marshmallows every evening and spray them with insecticide every morning. Joe and Bobo have huge horns, whereas Mary's are more petite. But beware—one of Mary's party tricks is to stand next to you and then swing her head upward; you suddenly feel a massive horn hurtling up inside your shorts, lifting you off the ground. Extricating yourself is surprisingly difficult. But it could be worse; when she was younger, Mary liked to lean on people (apparently that's how a Water Buffalo tells the size of another Water Buffalo). Thankfully she's a friendly buff, and smarter than most. If she wants to come into Greg's house, she knows to bend her head sideways so that her horns fit through the doorway. Once, while driving through the village, I made the mistake of stopping to chat with Mary. I didn't expect her to put her head through the open window to nuzzle my lap in a search for treats. Nor did I realize how difficult it is to dislodge buffalo horns from a steering wheel, or to push a buffalo's head back out through a car window.

Middle Point is a bird-watcher's delight. Iridescent Rainbow Bee-eaters line up on the telephone lines. Whistling Kites soar overhead; if it's feeding day at the croc farm across the road, hawks circle in their thousands. Corellas and Sulphur-crested Cockatoos screech abuse at each other. Magpie Geese honk mournfully as they fly overhead on their way to raid the mango orchard. Greater Bower Birds bounce around energetically in the oleander thicket. Striated Pardalotes and brilliantly colored Crimson Finches zip by. But reptiles are hard to find by day. Small dragon lizards bob their heads at each other on the lawns, and an occasional Black Whipsnake hurtles across the road. But until the sun sets, most reptiles are well hidden.

As the sun disappears below the horizon, Greg and I don thick shirts, long pants, boots, and baseball caps. Every inch of exposed skin is doused in insect repellent. And none of your namby-pamby, greenie, limp-wristed concoctions. Students from Sydney arrive with politically correct repellents, with labels boasting that they are "organic" and "environmentally friendly." One night on the dam wall has them racing into the Humpty Doo supermarket the next day to invest in tropical-strength DEET-based fluids. It's awful stuff (Greg's steering wheel is worn down in hand-shaped indents where the repellent has eroded it), but better than the Ross River fever transmitted by mosquito bites. Ask any long-term resident of the field station.

Our destination is the same every evening: the wall of Fogg Dam, where we can count wildlife and catch snakes as we walk along beside the water's edge. A sign prohibits walking on the dam wall, because of danger from crocodiles, but we have permission from the rangers. Greg uses a headlamp because he needs to have both hands free for snake grabbing. But my job is just to find snakes, not catch them, so I opt for a handheld flashlight that doesn't encourage the flight of insects toward my face. LED technology has revolutionized illumination in the bush; frog biologists discuss the latest LED models the way car enthusiasts talk about the newest V8 engine, or how fashionistas critique the season's frock styles.

The first part of our walk is lined with Pandanus Palms and Silver-leaved Paperbarks, and I always feel like I'm in a cathedral as I move through it. A group of my Ph.D. students came to a different view; they convinced each other that the dam wall is haunted, with unearthly presences gliding around. The only ones I encounter are mosquitoes, but fortunately their numbers are low during the dry season. Beetles cannon into us, attracted by our lights, then crawl around inside our shirts and (worse) underpants. Tonight, though, they aren't biters.

Owls flit from post to post ahead of us as we walk, but the first reptile we see is a Long-necked Turtle, striding purposefully away from the drying floodplain and toward the water. It's a hazardous journey because the dam wall is 3 meters (10 feet) wide, and turtles are not cheetahs. I soon overtake this miniature tank and admire its bizarre body form, with a neck that's longer than its backbone. Leaving it to continue its plodding progress, I scurry to catch up with Greg. He didn't bother to stop, even though turtles are his favorite animals.

It's pleasant strolling along the dam wall. Greg's fast pace keeps us ahead of the slower insects but makes it difficult for me to look carefully for snakes on the road edge. I also need to scan for crocs out in the dam, and avoid stepping on Death Adders, leaving little time to search the roadside vegetation for smaller serpents. My flashlight beam flicks around like a demented goblin, from water to dam wall to the base of a Pandanus Palm. I'd like to stop and look up into the foliage for frogs and tree snakes, but if I did, Greg would already be way ahead of me. For a short guy, he moves quickly.

Ten minutes after leaving the car, we see a snake crossing the dam wall ahead of us. And it's a big one—heavy bodied and almost 2 meters (more than 6 feet) long. But this iridescent "rainbow serpent" attracts only a snort of disgust from my companion. Water Pythons are beneath Greg's dignity. Too big, too flashy for his refined Canadian tastes. So we continue without molesting the big serpent. Whenever Greg sees one, he just calls out "Water Python" in a derogatory tone, as if he were saying, "Watch out, don't step in the dog poo over here."

Greg knows every inch of the dam wall, and before long he spots another snake. It's a Keelback—a slender, rough-scaled, nonvenomous species, closely related to the garter snakes of North America and the grass snakes of Europe. These inoffensive frog-eaters were the main focus of Greg's research world until the Cane Toad invasion arrived. He's individually marked thousands of Keelbacks, carefully clipping and cauterizing a unique combination of belly scales on each one. Strangely, Greg doesn't particularly like snakes. He only accepted the job I offered him, back in 1998, because he wanted to move somewhere warm.

Compared to the thrilling footage on "man versus snake" TV shows, Greg's capture method is prosaic. He just walks up slowly and pounces on the snake with a gloved hand. In the early years, he used a bright-blue rubber washing-up glove, soon lacerated by hundreds of snake teeth. A few years ago, he graduated to a snug leather version, but I still treasure the memory of Greg walking down the dam wall, surrounded by mosquitoes, wearing a bright-blue rubber glove. Most snakes ignore his approach until it's too late, but occasionally one takes off at top speed, forcing Greg to explode into action with a flying dive, to seize the snake before it vanishes into the water. Almost as soon as the snake is in his hand, it's been checked for scale-clips and then bundled into a small cloth bag with a number scrawled on it. Greg is walking again, even as he's entering the capture location in his field notebook.

A few minutes later, he pauses momentarily to check his notebook, before untying a bag and flicking a snake off into the safety of the reeds. It's a Keelback, about 50 centimeters long. He caught it last night, and measured and identified it this morning based on the unique combination of belly-scales he had clipped a few months ago, when the snake was a baby. With luck, the snake will turn up again in the future. Optimism is essential for any mark–recapture program.

The dam wall is 1.5 kilometers (a mile) long, and it takes us an hour to walk up and back. We see the bright gleams of crocodile eye-shines in the dam, and check their locations periodically in case any of them want to widen their menu choices. But they don't. The Fogg Dam crocs are well fed and have never shown much interest in people as prey. Sadly, the patriarch of the dam wall, a 4.5-meter (15-foot) male Saltwater Croc, was recently captured and removed by the rangers, because tourists complained about his presence. We're hoping that the Boss Croc will be replaced by one with an equally respectful attitude toward snake researchers.

The only croc that causes us any grief tonight is small, but she's a serial offender. She's a Freshwater Croc, a bit less than 2 meters (6.5 feet) long, who lays her eggs on the dam wall every year. Her latest clutch was laid a week ago; every night since then, she has maintained a vigil to deter interlopers. Unfortunately, her admirable maternal feelings aren't matched by any careful discrimination about who constitutes a threat to her eggs. Well-intentioned snake-hunters are attacked just as vigorously as nest-robbing Bandicoots. Greg knows all about this obsessive "freshie," but he doesn't mention to me that she has already nested this year. And he ensures that he is on the far side of the dam wall as we approach her nest. My first inkling of this scaly paragon of maternal devotion is when she hurtles up out of the water, jaws agape. Her hisses almost drown out the sound of Greg giggling. I retreat, with the croc in hot pursuit. I know she is bluffing, because last year I fell over when she charged, and she stopped long enough to let me recommence my cowardly retreat. But even if part of my brain knows she's bluffing, another part is screaming out that a crocodile is running at me with its mouth open. That second part sits deep within some primeval fear center in my brain. Crocodiles have been eating our ancestors for a long time.

The adrenalin keeps me focused as we walk back toward the car. Suppressing the urge to throw Greg into the dam, I see another snake—thinner than a Water Python, but more muscular than a Keelback. It's a Slaty-Gray Snake, Greg's other study species. Unlike Keelbacks, Slaty-Grays are homebodies; each individual stays in a small area on the dam wall instead of traveling widely across the floodplain. So Greg recatches many Slaty-Grays time after time, and over a long period—unlike Keelbacks, which only live for a year or two, Slaty-Grays can survive for at least a decade. As a result, Greg's data on Slaty-Grays are very different from his data on Keelbacks—many captures of a few individuals versus a few captures of many individuals. That lets us explore different questions. At the moment, for example, Greg is looking at the inheritance of immune responses. Because he caught and blood-sampled the parents and grandparents of today's Slaty-Gray Snakes, he can identify the fathers of the young snakes on the basis of genetic paternity tests, and look at how immunology runs in families.

The night's tally is three Keelbacks and two Slaty-Grays. The five Water Pythons also get noted down, although we didn't bother to catch them. A few Rocket Frogs jumped across the dam wall in gigantic bounds, but none ended up in Greg's survey quadrats. One Golden Tree Snake sleeping up in a Pandanus Palm midway across the dam wall, a Bandicoot, and a White-tailed Water Rat round out the information jotted down in Greg's sweat-stained notebook. It doesn't mean much on its own, but when combined with thousands of other nights of data, it tells us what's happening on the floodplain.

Greg and I head back for dinner. Houseguests are rare in Middle Point (especially if you live with Greg), and Cath prepares wonderful feasts whenever I visit. Curry tonight, one of my favorites. As we pull up to the house, I hear the sound of a food processor at work. Frozen mangoes are being crushed to create a daiquiri. Life is good. I record the night's tally and experiences in my field diary, then head down for a predinner drink. Greg lies on the sofa in his bathrobe, holding a piece of mincemeat for his pet Bluetongue Lizard. Lucky is undecided whether to consume the meat or Greg's toe—disappointingly, she settles for the meat. Cath looks on tolerantly.

Greg Brown with Joe, one of his pet Water Buffaloes. Photo by Terri Shine.

The days blend into each other. In the office by day, talking to students and answering emails; back to Greg's house for lunch; wash last night's dishes; lie down for twenty minutes. And then back to the office, and more talking and more tapping at the keyboard; and right at dusk, out the door to walk the dam wall yet again.

By 2000, though, it became clear that my idyllic world of snake research at Fogg Dam was in peril. Cane Toads were transforming from a distant threat into a looming catastrophe. In 1983 they crossed the border from Queensland into the Northern Territory, and by 2002 they were killing reptiles in Kakadu National Park. By that time Thomas had returned to Sweden, and Greg was the only full-time researcher at Middle Point. Funding for "blue sky" science was more and more difficult to obtain. I belatedly realized that the toads would reach Fogg Dam within a few years—and when they did, they might spell the end of snake research at Middle Point.

Why hadn't I anticipated the toad invasion much sooner? I knew full well that the tidal wave of alien amphibians was heading toward Fogg Dam, but I had underestimated its speed. Based on the rate of toad invasion in 1985, government researcher Bill Freeland had predicted that the aliens would arrive at Middle Point in 2027. In fact, it took them half that time. For reasons I'll talk about later, the toads accelerated as they headed west.

I had two reactions to the imminent incursion. On an emotional level, I hated Cane Toads for defiling the ecosystems I loved. But as a scientist, I couldn't ignore the one-in-a-million chance to study a biological invasion as it swept through an area where we already had extensive baseline data. Fate had dealt me a unique set of cards.

It would be unethical as well as idiotic to ignore that opportunity. Via the government grants system, the Australian public had funded me to study the snakes of Fogg Dam for many years. Now there was a chance to value-add to everything that we had achieved, by switching across to a problem that Australians really cared about. The residents of Darwin and Humpty Doo were expecting ecological meltdown from the toad invasion. I had been handed an extraordinary opportunity to unravel the truth about that invasion—and perhaps even work out what we could do about it.

As the invasion approached Darwin, the media increasingly demonized the Cane Toad and predicted ecological calamity. This exaggerated rhetoric ramped up public concern even higher—and, ironically, helped convince granting bodies that Cane Toads were a major problem. My next grant application was about Cane Toads, not snakes. It worked. In 2003, the Australian Research Council gave me the largest grant I had ever received—a fellowship to pay my salary, plus enough cash to rent most of the houses in Middle Point Village, and to expand my research team. It was just in time.

As the invasion moved through Kakadu National Park and toward Fogg Dam in 2004, the nucleus of a toad research team gathered in Greg's house on Beatrice Hill Farm. We hadn't yet taken over Middle Point Village, and Greg was living in a demountable building on a Water Buffalo farm a few kilometers away. The buffalo boys were very supportive of our work. The manager of the farm, Ken Levey, was a large, overweight man—at first sight, the stereotype of a redneck rural Territorian. But appearances can be deceptive; that burly frame concealed an extraordinary intellect. If I wanted to know something about Etruscan art, or Swedish-designed military planes, Ken was the man to ask. Although his job was all about moving buffaloes from one paddock to another, the man was a walking encyclopedia. He never made a public show of his knowledge; it was revealed only if you asked him a question or (after he knew you well enough) if you made a mistake in some supposedly factual statement. Memorably, Ken once corrected me gently about the publication date of Darwin's *Origin of Species*. And of course he was right. Meeting Ken was a crash course in the folly of assuming that professors are smarter than buffalo farmers.

Almost as soon as we began working with toads rather than snakes, I invented a name for my research group: "Team Bufo." We wouldn't be able to escape the media frenzy around toads, so we might as well get some branding in place. Other groups had already taken the obvious terms like "toadbusters"—and anyway, we were trying to understand toads, not bust them. In a public talk I gave at Sydney University in 2005, I included an Acknowledgments slide listing the names of my postdocs and students, to credit them for their work. The page needed a heading—so I cast about for a light-hearted title. The night before, my family and I had watched the comedy film *The Life Aquatic with Steve Zissou*, featuring the scuba-divers of "Team Zissou." The name "Team Bufo" sprang to mind. A journalist highlighted the name in a media story, and it took on a life of its own. I embraced it, and even had "Team Bufo" bumper stickers made up for our field vehicles.

The first discussions about toad research involved just four of us—the professor (me), two postdocs (Ben and Greg), and a student (Matt Greenlees). Matt is a genial, larger-than-life bear of a man, with a gold earring and an unquenchable passion for life. Beer, snakes, and frogs were his chief delights until he settled into domestic bliss. His wife and kids now have the top spots in his attachment hierarchy, although he still slips away to wander around swamps at night whenever he can. Matt was often the comedian as we sprawled on the floor (due to a shortage of chairs) in the small living area of Greg's demountable, trying to work out how best to get the toad project underway.

We decided to focus on the ecological impact of Cane Toads, and to do it as broadly as possible. Our top priority was the native species most likely to be affected: the frogs. Keen to do the first toad-focused project at Middle Point, Matt centered his undergraduate (honors) research in 2005 on the impacts of toads on other amphibians.

But we couldn't stop at frogs. Toads might affect all kinds of other animals as well—but which ones? It was frightening. I was out of my depth, and more-than-usually receptive to ideas from my colleagues. We were poised at the threshold of an extraordinary opportunity. We had phenomenal baseline data from Greg's nightly surveys, and within a year we would be overrun by a high-profile pest expected to wreak havoc on our study animals. Despite predictions of doom, nobody really knew what effects the toads would have.

My fear came from a simple issue: we only had one opportunity to measure the ecological impact of the Cane Toad. We needed to work out what to do, and we needed to start doing it *before* the invaders arrived. We could never repeat the Fogg Dam studies anywhere else, because we wouldn't have the background data. I lay awake at night, frantically trying to identify any gaps in our plans. If we picked the wrong species to study, or failed to collect critical "pre-toad" data, there would be no second chance.

Snakes were the center of my research world, so I knew how to look at their responses to the Cane Toad onslaught. But what about the effects of toads on fishes, birds, lizards, and marsupials? We couldn't do it all ourselves, so I decided to recruit students to work on those topics. By the time the students realized that their supervisor was hopelessly ignorant about everything except snakes, it would be too late for them to withdraw. Matt took on the frogs, Christa Beckmann looked at birds, Ruchira Somaweera at crocodiles, Stephanie O'Donnell at quolls, Sam Price-Rees at Bluetongue Lizards, David Nelson at fishes, Georgia Ward-Fear at ants, Elisa Cabrera-Guzman at other insects, and so forth. My longtime collaborator Jonno Webb ran the marsupial side of things. We couldn't cover every native species, but we filled in many of the remaining gaps with targeted studies. By 2004, as toads approached Fogg Dam, the stage was set: a small group of researchers who knew a lot about snakes and the local environment, but were woefully ignorant about amphibians, versus an oncoming wave of gigantic toads from South America.

To an outsider, the difference between studying snakes and studying frogs looks trivial. But in fact, the two types of animals differ in important ways. Adult Cane Toads weren't a problem: they wander around in the same habitats as snakes, so I could use similar techniques to study them. I would have to attach a toad's radio transmitter to a waist belt, instead of surgically (as in snakes)—but once I'd released the animal back into the bush, I could follow that signal and relocate my animal just as I'd been doing with snakes for many years. Ditto for ways to measure and analyze movements, home-range sizes, dietary composition, and the like. So the snake biologists of Team Bufo should fare well in studying adult Cane Toads.

But frogs and toads have an extra dimension to their life history that doesn't occur in snakes. Cane Toads begin their lives as aquatic eggs, then spend a month or two as tiny black tadpoles before metamorphosing into juveniles capable of living on land. And the mysteries of tadpole life were far beyond my ken; these little wriggly beasts might as well have been space aliens. I began to wish that I'd paid more attention in herpetology conferences, instead of leaving the room when anyone began to talk about tadpole ecology.

We could never understand Cane Toad ecology unless we could grasp the subtleties of the tadpole stage, so I needed someone to fill that gap in Team Bufo's skill set. And as with so many things in those early days, the right person arrived at the right time. I had enough funding to employ one more postdoc, and I found the perfect Larval Guru. Michael Crossland is a tall, slender, soft-spoken Queenslander with a passion for tadpoles. Michael can look a tadpole in the eye and know what it's thinking. If a tadpole has a headache, Michael feels its pain. And unlike snakes, tadpoles are small enough and numerous enough that the best way to study them is to run large-scale experiments. That takes a precise mind—and in contrast to my own inability to focus on details, Michael revels in them. The thought of setting out hundreds of replicated containers for an experiment makes him smile, whereas it makes me run for cover.

I was lucky to find him. Apart from a few zealots like Michael, researchers who study amphibians focus on adult frogs, not their larvae. Most tadpole researchers are unusual characters, because it takes a special kind of person to devote their life to these little beasts. The rest of us find it easier to get excited about colorful adult frogs, rather than delve into the mysteries beneath the murky surface of the pond. At an amphibian biology conference, the tadpole enthusiasts seek each other out at the opening social event—and bore everyone else witless by swapping tadpole pictures. If you think human babies all look the same, you should check out tadpoles. But to a true-blue tadpole biologist, a frog is just a tadpole's way of making another tadpole.

As befits a details devotee, Michael has strong likes and dislikes. He brings a careful discrimination to all parts of his life, not just his work. For example, Michael has a passion for Easter-egg chocolate rather than "normal" chocolate—but even then, not all types of Easter-egg chocolate are acceptable to his sophisticated palate. Easter comes just once a year, so Michael plans ahead. Ever since I hired Michael, every fridge and freezer in the lab has filled up before Easter each year. They are crammed with 400-gram Red Tulip Easter Bunnies, to be gradually consumed over the next several months. Many a student searching for a frozen toad has been sorely tempted as they fought their way past serried ranks of frozen chocolate rabbits to locate their target. When the rabbit armada runs out, cartons of Peppermint Patties or dark-chocolate Milky Way bars move in to fill the niche left vacant by the Easter Bunnies.

Team Bufo girded its virginal loins for the approaching battle. Greg assembled the equipment he would need to radio-track toads across the floodplain: quad bikes, transmitters, receivers. Michael purchased trestle tables for the lab, and thousands of small plastic takeaway food containers to put on those tables, to hold tadpoles. He could use those arrays to ask tadpoles simple questions—such as whether or not they will try to eat toad eggs if given the opportunity, or how the presence of one type or size of tadpole affects the growth or survival of others. But exactly what other questions should we ask? That took a lot of discussion.

We were flying by the seat of our pants, and our planning was based on intuition rather than knowledge of toad biology. But, to my delight, our venture into toad science attracted support the way a cowpat attracts dung beetles. Expert amphibian ecologists forgave my blunders as I tried to understand these new (and, to me, very perplexing) animals. Nobody made jokes about stupid snake biologists who didn't know one end of a frog from the other (or at least they made those jokes when I wasn't in the room). Instead, they warmly welcomed a recycled snake biologist into the world of frog biology. The leading lights of Australian amphibian ecology enthusiastically collaborated with us.

Local people in the Darwin area were supportive also—and that's important. The locals know the landscape far better than any scientist. The list of supporters included Darwin-based biologists as well as farm managers, Parks and Wildlife rangers, and the Northern Territory government employees who leased out the village to us. But one character in particular played a seminal role. When I first began to study Water Pythons at Fogg Dam in the 1980s, a young farmworker named Eric Cox took an immediate interest. He quizzed me in such detail about our activities that I began to suspect that he was a spy for the wildlife authorities, checking if we were smuggling snakes back to Sydney for the pet trade. But as I came to know Eric better, I realized that he was simply fascinated by wildlife and determined to help us in every way he could. Many years later, I had the pleasure of nominating Eric for a prestigious national award—an Order of Australia—to thank him for his support. He's the only buffalo farmer to ever win that award.

At the beginning, I viewed Cane Toads in black-and-white terms, as a blight on the Australian landscape. I had to study them, but I didn't have to like them. Cane Toads were the enemy, and our research project was a thinly disguised crusade.

Although I was born in Queensland, where Cane Toads were introduced long before I was born, I had little firsthand contact with the alien amphibians until 2005. My family left Brisbane when I was five years old, and the toad tsunami didn't arrive in Brisbane until twelve years later. My formative years were spent in Melbourne, Sydney, and Canberra, places too chilly for these tropical amphibians. So I didn't have much to do with toads when I was a youngster. I laughed when people joked about hitting toads with golf clubs or splattering them on the roads. I saw Cane Toads during family holidays to the southern Queensland coast, but I didn't give them much thought. During fieldwork on Fraser Island in 1970, a forestry worker showed me a grainy photograph of a dead Death Adder beside an equally dead Cane Toad. I was horrified.

Many years later, toads came back into my world when my aging parents retired to southern Queensland in the early 1980s. Mum and Dad took regular evening walks around the local streets for the first few years—until they discovered that Cane Toads roamed the suburb after dark. My parents were trapped inside their house at night by the specter of giant amphibians, for which they felt nothing but revulsion, on their lawn. One of my father's last acts, before he collapsed in the backyard with a heart attack, was to skewer a Cane Toad with his garden fork. When I flew up for Dad's funeral, I found the toad impaled on the tines of the fork, still very much alive. It was my first experience of the extraordinary tenacity of these animals. Dad would have wanted me to kill the toad, but the thought of another death was just too awful. I released the toad and let it hop away.

Professionally, Cane Toads were not an issue for me for the first forty years of my career. When I thought about them, I took the politically correct line. In my lectures to undergraduates, I warned that Cane Toads were a horrible pestilence. In a radio interview, I likened the imminent toad invasion to "hitting the ecosystem over the back of the head with a shovel." The toad-busters would have applauded my apocalyptic view.

But my fellow members of Team Bufo didn't agree with that anti-toad zealotry. Many people advocate "respect for all life," but very few manage to live it. Ben Phillips does, and it's difficult to be casually brutal when there is a conscience in the room. To Ben, Cane Toads were living creatures rather than evil demons. There's more than a hint of hippie-ness about Ben. Not the extremism of a tie-died, zen-master hippie who believes that every bowel motion is a spiritual experience, but a mellow thoughtfulness. Ben sidesteps socially fabricated blinkers. If you need anyone to point out clearly, succinctly, and unambiguously that the emperor has no clothes (and damn the consequences), Ben is your man.

Greg Brown's attitudes were challenging also. Greg was at the cutting edge of the toad program at Middle Point; he was the one who was driving the roads every night and strapping radio transmitters to the first toads. Much of what I learned about toads came through Greg's discoveries. And, inevitably, that was filtered through the perspectives of a man who prefers the company of turtles to that of people. Ben may not pander to social norms, but Greg takes it further—he doesn't even notice them. The media blitzkrieg about the impending catastrophe, and the demonization of Cane Toads, washed past Greg without affecting his opinions. When the invading toads arrived, Greg embraced them with open arms. He likes toads and always has. As a child in Canada, his nickname was "Toad." So while I was working through politically correct condemnation of the loathsome invaders, Greg was looking forward to having some new friends to play with.

And so we were ready for the invasion—and in the nick of time. After the pioneer members of Team Bufo assembled at Middle Point, we often drove at night along the Arnhem Highway east toward Kakadu National Park. Within an hour, we could find the invasion vanguard of Cane Toads hopping along the road. A few months later, we only had to drive for thirty minutes before the road was covered with Cane Toads. And finally they arrived. Just before Christmas 2004, Ben saw an adult female Cane Toad hopping on the road ahead of him near Middle Point. He caught the toad and toe-clipped it so that we could identify it later (losing a toe or two doesn't worry a toad, or even elevate its stress-hormone levels). Then he released it where he found it. It was a true pioneer, well in advance of its fellows. We didn't see any more toads for two months, until late one night when Greg encountered that same toad even closer to Middle Point. This time, Greg strapped a radio transmitter on her to follow her movements.

From then on, a trickle of toads appeared in our study sites, to be duly marked, measured, and equipped with transmitters. Within a few months, the trickle turned into a flood. Flattened toad corpses dotted the roads. As dusk fell, toads emerged from beneath the houses and sheds at Middle Point and sat under the lights to catch insects. The Cane Toad invasion had arrived, and Team Bufo swung into action.

FOUR

How Cane Toads Have Adapted and Dispersed

> Nothing in biology makes sense except in the light of evolution.
> **THEODOSIUS DOBZHANSKY**, *American Biology Teacher*, March 1973

As we humans travel around the planet, we often bring animals and plants to places where they don't belong. Sometimes it's accidental (like strains of flu spreading across the world), but many translocations are intentional. Sadly, for example, European settlers tried to "improve" the Australian landscape by introducing creatures from their homelands. Confronted with the alien Australian bush, homesick Europeans longed to see familiar plants and animals in the backyard. The result was ecological catastrophe.

Fortunately, the result isn't as devastating on a large continent as it is on a small island. New Zealand is an ecological Chamber of Horrors, with a few native species clinging to existence in a world full of interlopers. The same is true in many other island nations. But, although the vast expanses of Australia allowed many natives to persist, the intruders spread rapidly. Our cities harbor Indian Mynah Birds, Black Rats, and Maple Trees. Our pastures are covered by Scotch Thistles. Gamba Grass from Africa fuels torrid fires that kill even the hardy eucalypts. Our rivers are awash with waterweeds that were brought in to beautify fish tanks. The roots of Willow Trees break down river banks, and our rivers teem with Carp. Rabbits destroy plant life across southern Australia, Camels chew through the fragile flora of the desert, and Water Buffaloes munch their way through tropical floodplains. Feral predators like cats and Foxes have sculpted Australia's tragic history of marsupial extinctions. Over the past two hundred years, half of the mammal species that have gone extinct worldwide were Australian.

The success of many invaders is no cause for surprise. We might expect Camels to thrive in the Australian desert, and sophisticated predators like cats and Foxes can make a living almost anywhere. But the same can't be said for Cane Toads. These large, ungainly amphibians don't look like storm troopers. They seem too slow, dim-witted, and inflexible to be successful invaders, and too dependent on water to conquer the harsh Australian outback. Clearly, appearances are deceptive. The Toad Army has swept through tropical Australia like a small, warty version of the Nazi tanks that poured into Poland in 1939.

But as the battalion of toads moved inexorably toward Darwin, they were confronted with a new enemy: the recycled snake biologists of Team Bufo. With our expanded facilities and bigger budgets, by 2005 Middle Point had already become the largest center of amphibian-ecology research in Australia. That's not as impressive as it sounds, because funding for frog research is minimal. The high public profile of Cane Toads meant that we could afford to invest in the infrastructure we needed. We could run large-scale trials and test exciting but risky ideas. Working on an abundant species and with enough money to do it right, gathering data was easy. In comparison to studying snakes (rare, secretive, uncooperative animals), doing research on Cane Toads was a breeze.

Team Bufo's first major discovery was about the speed of toad invasion, and the reasons why that speed has changed. As I mentioned in chapter 1, toads are world travelers. As the ancestors of today's toads circumnavigated the planet over millions of years, the best dispersers were large species that could tolerate dry conditions, had big poison glands, and laid vast numbers of eggs. The Cane Toad fits that stereotype. They don't come much bigger, tougher, more toxic, or more fecund than *Bufo marinus*. So Reg Mungomery and his mates at the Sugar Experiment Station unwittingly chose a toad species whose forefathers had specialized in long-distance travel. Small wonder, then, that Cane Toads took off for the farthest corners of Australia as soon as they had the chance.

Long-distance travel isn't a common feature of amphibian life. Most frogs are homebodies, remaining in a small area rather than heading off across a continent. But toads (of all species) are special. Their physiology is well suited to long-distance excursions, and they are mobile even when they are living at home rather than invading a continent. For example, European Toads often travel many kilometers between their winter burrows and the ponds where they breed in spring. Even in places where Cane Toads are fairly sedentary—like Venezuela, Hawaii, and eastern Australia—radio tracking usually reveals at least one or two long-distance movements (often for no apparent reason). The Cane Toad was born under a wandering star.

Until Cane Toads were brought to Australia, nobody in the toad family, the Bufonidae, had the opportunity to express the full extent of that wanderlust. Nonetheless, a toad with a passion for travel has a central role in one masterpiece of children's literature, about woodland creatures in the English countryside. This fast-moving amphibian hero is Mr. Toad from Kenneth Grahame's 1908 classic tale *The Wind in the Willows*. Mr. Toad's obsession with speed ultimately leads him to steal an unattended motor car, with predictable results:

> The car stood in the middle of the yard, quite unattended, the stable-helps and other hangers-on being all at their dinner. Toad walked slowly round it, inspecting, criticising, musing deeply.
>
> 'I wonder,' he said to himself presently, 'I wonder if this sort of car STARTS easily?'
>
> Next moment, hardly knowing how it came about, he found he had hold of the handle and was turning it. As the familiar sound broke forth, the old passion seized on Toad and completely mastered him, body and soul. As if in a dream he found himself, somehow, seated in the driver's seat; as if in a dream, he pulled the lever and swung the car round the yard and out through the archway; and, as if in a dream, all sense of right and wrong, all fear of obvious consequences, seemed temporarily suspended. He increased his pace, and as the car devoured the street and leapt forth on the high road through the open country, he was only conscious that he was Toad once more, Toad at his best and highest, Toad the terror, the traffic-queller, the Lord of the lone trail, before whom all must give way or be smitten into nothingness and everlasting night. He chanted as he flew, and the car responded with sonorous drone; the miles were eaten up under him as he sped he knew not whither, fulfilling his instincts, living his hour, reckless of what might come to him.

The Cane Toad's advance across Australia has been documented in more detail than almost any other invasion worldwide. Because of the Cane Toad's distinctive appearance and its fondness for human habitation, people *notice* when toads first arrive. Far-sighted scientists conducted surveys to capture that knowledge. For example, Eric van Buerden and Gordon Grigg sent out postcards to schools across Queensland, asking when toads had reached each area. Later, other researchers like Wendy Seabrook, Simon Easteal, and Mike

Sabath did the same. As a result, we have a clear picture of where the toad invasion front moved, and how quickly. In comparison, for example, we have only a rough idea of how Rabbits and Foxes spread through Australia, because it happened too long ago, and nobody was paying attention.

So, just how fast have Cane Toads moved across Australia? The early records show a steady progress of 10 to 15 kilometers (6 to 9 miles) per year. For a small animal, that's impressive. A toad that travels in a straight line, as fast as it can go, covers about two-and-a-half body lengths in a single hop. So, even if a toad set out along the road, without any meandering, it would require at least 40,000 hops to move 10 kilometers (6 miles). And, of course, no toad moves in a straight line.

Remarkably, though, that was just the beginning. As Cane Toads traveled across Queensland and then the Northern Territory, the toad front line moved faster and faster. By the time they got to Kakadu National Park in 2002, toads were spreading at 50 kilometers (30 miles) per annum—despite moving through a seasonally dry region, where they can only travel cross-country for a few months each year. On wet-season nights when the soil is moist, the toads really get moving. The first toads to arrive at Fogg Dam were extraordinary athletes, traveling a kilometer (half a mile) every night. In contrast, the toads that our colleagues Ross Alford and Lin Schwarzkopf had tracked in Queensland moved less than 10 meters (11 yards) in a night. Compared to the Queensland couch potatoes, our invasion-front toads were long-distance runners.

How did the Middle Point toads achieve these incredible distances? Greg's radio tracking revealed several tricks. First, the invasion-vanguard toads moved all night, not stopping to rest until the sun came up the next morning. Second, they abstained from sex: although there were occasional pairings in roadside ponds, most of the invasion-front toads kept sprinting westward rather than breeding. That saved them time and energy to devote to travel. Third, they picked the easiest path—along the road rather than through the bushland. Fourth, they switched feeding tactics; instead of waiting in one place to ambush prey, they grabbed insects on the road as they went—the fast-food option. And fifth, the toads changed the way they moved. Long leaps are good for escaping a predator, but small hops are a more efficient tactic if you need to keep on going for hours. Use your forelegs as well as your hindlimbs, so that the elastic recoil from your last hop powers the next one. And, last but not least, night after night the invasion-front animals kept moving in the same direction—northwest—rather than meandering like their Queensland ancestors.

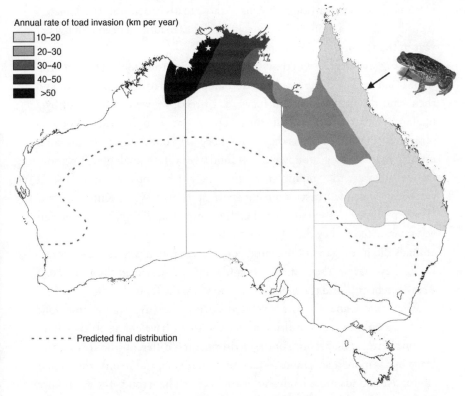

Cane Toads were introduced to northeastern Australia (Queensland) in 1935 (toad symbol and arrow indicate location). They have since spread west and south. The map shows the rates at which toads have spread, and the dotted line shows the predicted final distribution of the species in Australia. The star shows the location of our research station at Middle Point.

After all of our careful preparation before the toad front arrived, we were astonished (and horrified) by this athletic behavior. When we sat around in Greg's demountable in 2004, planning the radio-tracking studies, it seemed simple: catch the first toads to arrive, pop transmitters onto them, then locate the toads every day to see how far they move. But we based our thinking on slow-moving Queensland toads, expecting them to stay in the same area for weeks at a time—not sprint away as fast as their stubby little legs could carry them. Even though Greg knows the area around Fogg Dam intimately, many toads moved too fast for him to keep up. They headed off toward Darwin, carrying their expensive radio transmitters with them. We got a few of our toads back when they wandered into backyards in Humpty Doo and the locals noticed their transmitters. One of those toads had moved 21 kilometers (13 miles) in only a month—about twice as far as toads had been moving per year in the 1930s, after they were first released.

The biggest surprise was that the Cane Toad invasion looked like an invasion. I had expected the newly arrived amphibians to settle in, breed, and then send the next generation forward. Instead, we were confronted by a phalanx of battle-hardened storm troopers—large adult toads—hopping obsessively onward as fast as they could. I began to understand why the community toad-busting groups were so fond of military analogies. Invasion-front toads were the amphibian equivalent of the troops sent in to assault vigorously defended castles during the Napoleonic Wars. Known as the "Forlorn Hope," those soldiers had little chance of surviving, but promotion was guaranteed if they did. And so it is for the toads—many invasion-front animals die in the jaws of predators, or shrivel up in a waterless landscape. But if they survive, they inherit a bug-filled paradise—until the main Toad Army catches up, and the pioneers have to share the food.

Why have Cane Toads accelerated during their Australian invasion? There are only two possibilities: either the toads have changed (they have become better at dispersing) or the environment is different (the area around Darwin has more roads, more food, more wet nights, and so on). The second alternative sounds more likely, but it turns out to be wrong. The toads have moved faster and faster through all kinds of habitats, not just the warm, wet, bug-rich swamps of the Northern Territory. Can it be the toads themselves that have changed? To test that idea, you can catch toads from different places (say, from Cairns versus Darwin), and release them in the same place at the same time (with radio transmitters attached, so you can follow them around). Ben conducted exactly that study at Fogg Dam, and the results were clear. The invasion-front toads moved much farther than their Queensland counterparts. Toads have changed.

But why? That change in toad behavior isn't necessarily "evolution." It might be nurture, not nature. Perhaps the environment around Darwin causes a toad to grow up as a racehorse and not a slowpoke? If so, this wouldn't be inherited—every generation would start again from the beginning. For the change in invasion speed to be due to evolution, the traits that accelerate dispersal need to be inherited from parent to progeny.

In most study systems, we could never test that possibility. For example, I couldn't imagine breeding snakes in captivity to see how characteristics evolve. But Cane Toads are abundant, have large clutch sizes, and have diverged dramatically within eighty years, so perhaps it was possible to go a step further and see if the change in Cane Toads as they poured across tropical Australia was a genuine—and spectacularly rapid—example of evolution in action.

I began to entertain a heretical idea: that studies on Cane Toads might shed light on the process of evolution itself. We had stumbled across a superb "model system" to look at evolution as it happens. Rather than just documenting ecological carnage, Team Bufo might be able to ask fundamental questions about the processes that sculpt biological diversity.

But to ask those questions, we had to understand whether or not specific traits were heritable—because if they weren't, any changes we saw would likely be due to environmental effects, not evolution. The only way to assess heritability was to breed toads on an industrial scale. In 2006, Ben collected adult toads from across Australia (from their original release site in Cairns across to the Northern Territory invasion front), bred them at Middle Point, and then raised their offspring under identical conditions. Two years later, when the young toads had grown large enough to carry radio transmitters, he released and radio-tracked them. It was an ambitious project, but it would tell us whether the change in Cane Toads from east to west was due to environmental effects or to evolution.

I didn't say too much about this side of our work in public, because it was a political minefield. At the same time that the toad-busters were using government funding to slaughter toads, we were using government funding to breed more toads. If the local newspapers found out about the project, the next day's headline would be "Boffins Undo Toad-Busters' Good Work." And at first, I wasn't even sure that it would be possible to breed toads from different parts of Australia and raise their progeny to adulthood. It's a quantum leap above maintaining adult toads in captivity. If you want your captives to breed, you have to give them all the comforts of home. And as every parent knows, raising even one baby is a demanding task. Raising a few thousand would be challenging.

The success of Ben's project to breed and raise toads was a testament not only to his dedication, but also to the continuing expansion of facilities at Middle Point. I employed a research assistant, Michelle Franklin, to take over the routine husbandry tasks, thereby freeing Greg, Ben, and Michael from the day-to-day drudgery. Michelle's first job was to organize the construction of a large metallic sign reading "Tropical Ecology Research Facility," with the Sydney University logo. We proudly erected it beside the road leading into the village. Constructing the sign was much easier than obtaining permission to do so. I didn't care what we called the facility; I just wanted to let tourists know that the local Parks and Wildlife office had moved elsewhere. But the university couldn't decide whether we should be called a "Centre" or a "Facility." I finally went ahead and ordered a sign anyway. Nobody complained, and it's a damn attractive sign.

One by one the initial teething problems were solved, and the village was transformed from a ranger station into a research station. One problem became apparent as soon as we began raising tadpoles, though. They all died. Our water supply from a nearby well was chlorinated because it served the local school as well as the village. Chlorine is deadly to tadpoles. But if we stopped adding the chemical, the schoolchildren would succumb to raging diarrhea. We needed a separate water supply. Fortunately, Eric the buffalo farmer knew where the waterlines ran underground, and we hired a plumber to come out from Darwin and connect us directly to the well. Before long we had our own water tank beside the workshop, full of uncontaminated water. It was a great success from the tadpoles' point of view, though some of my students developed diarrhea.

Water was just the first of our worries. We also had to work out how to breed toads. We discovered that we could make the adults shed eggs and sperm by injecting them with amphibian hormones (the high-tech equivalent of the old medical method of injecting human urine into toads for pregnancy testing). And then we put our injected male and female toads together, in carefully selected pairs, and hoped for eggs. It was hit-and-miss in the early stages, but Ben gradually refined the procedures. Toads spawn at night, so Ben had to wake up several times a night to check the romantic duo. If spawning was in progress, he rapidly switched males so we could obtain baby toads with the same mother but different fathers: very useful for genetic analyses.

And that's when the workload really accelerated. As soon as the toads bred, we had thousands of hungry tadpoles to feed. Michael froze heads of lettuce so that they broke down into green mush as they thawed, easier for the tadpoles to chew. The piles of thawing lettuce heads on the floor of the workshop triggered a surprising addiction in Greg's dog, Slim Dusty. He couldn't resist the crunch of frozen lettuce in his jaws. Slim was a huge black hound who looked as if he'd be at home fighting a wild boar to the death. But in fact he was a gentle, lugubrious creature, and crunchy lettuce was his favorite snack.

Tadpole husbandry requires an organized mind. To keep track of who is who, you have to raise each tadpole separately. The workload is staggering, with thousands of takeaway food containers that need to have their water changed every day. And it gets even worse in the next stage: hordes of juvenile toads that need vast amounts of tiny insects for food. The local termite mounds took a real hammering, but I still spent tens of thousands of dollars a year buying baby crickets. Fortunately, Greg taught the adult toads to eat dog food, cheaper and more nutritious than crickets.

When the young toads were small, we housed them in plastic boxes. As they grew, we transferred them into large plastic crates, about a meter (3 feet) square. Designed to hold fruits and vegetables, they made perfect waterproof homes for growing toads. For the adults, we constructed communal outdoor enclosures made of vertical metal sheets dug into the ground, forming rectangular pens that measured 4 × 10 meters (13 × 32 feet). Electric lights strung on posts across the middle attracted bugs at night, providing abundant natural food. Each enclosure had a small pond, filled with unchlorinated water straight from the well. Fortunately, there were no Great Escapes—security was tight, and as soon as a toad was no longer needed, it was humanely euthanized.

We spent money where we had to—mostly on field vehicles and on husbandry equipment for Cane Toads—and saved it wherever we could (by skimping on upgrades to the offices, lab, workshop, and houses). So our facilities were Spartan. The main office–lab complex ("Toad Hall") consisted of a large lab, half a dozen small offices, and a big library-cum-tearoom. Right beside the main building, the large outdoor workshop housed Michael's tadpole experiments. Most of the workshop floor was covered by tables, supporting neat rows of small takeaway food containers that each held a single tadpole to expose it to some specific treatment. Water dishes under each foot of each table kept marauding ants away.

The lab (a big, open room with a large bench in the middle) was transformed into a toad research facility by bringing in a few microscopes and balances. Most of our research was low-tech, so we didn't need expensive instruments. Our only involvement with high-tech science came via Greg's interest in immunology. In one small room, he set up a gleaming machine labeled "absorbance/luminescence/fluorescence plate reader."

The offices were designed for one person each, but we crammed two desks into each room. As the first research-based inhabitants of the village, Greg and Ben claimed the biggest office. I attached a pompous sign to their door proclaiming "Senior Administrative Officer of the Institute, Dr Gregory P. Brown," and a smaller sign bearing Ben's name. After Ben left Middle Point, Greg moved his nameplate to the toilet door.

The only other change I made to the central office block was to affix a few plastic warning signs. Outside each external door, I screwed a plaque on the wall reading "Venomous snakes kept in this building. Enter at your own risk." I made the signs to discourage thieves, and it worked. Another engraved plastic sign went on the tearoom door, to officially dub it "the Mungomery Library" in honor of Reg Mungomery, the man who brought Cane Toads to Australia. If not for Reg's terrible mistake, there would be no toad research at Middle Point. It's also a reminder that scientists got us into this mess, and that we need to do better this time around.

Our facilities for doing science were primitive but effective. Ben's baby toads survived and grew. Within a couple of years, they were large enough for Ben to strap radio transmitters onto them and measure their dispersal rates, just as he had for their parents. The results were spectacular. The babies of invasion-front toads inherited their parents' fast dispersal, whereas the offspring of Queensland toads just sat around.

I was gobsmacked. Not only have Cane Toads changed as they colonized Australia, but those changes have a genetic basis. It's evolution, though not as we usually know it. Within a single human lifetime, a stay-at-home amphibian from Latin America has been transformed into a creature that is now invading several times faster than when it first arrived in the Great Southern Land. No matter how you look at it, Cane Toads have evolved dramatically in the course of their romp through Australia; the invasion-front individuals look different, they act different, and they pass these characteristics on to their offspring. No other vertebrate species, worldwide, has evolved so quickly and dramatically.

Further analyses of those trials showed that the young toads had inherited not only overall dispersal rates from their mother and father, but also some specific behaviors. For example, baby toads inherit their parents' "path straightness": if the parents kept moving in the same direction night after night, so do their progeny. If the adults meandered, moving west one night and east the next, their offspring do the same. The invasion-front toads aren't genetically programmed to head in any particular compass direction, but (unlike Queensland toads) they *are* programmed to keep moving in the same direction night after night. It's the most efficient way to move long distances.

Another key piece of evidence came from Greg's radio tracking. As the years passed, movement patterns began to change. After the frenetic one-way traffic of the first couple of years, successive cohorts of toads calmed down. They kept moving through, but within a few years the newly arriving amphibians were moving 100 meters (109 yards) in a night rather than a kilometer (more than half a mile). They didn't stay in "dispersal mode" as long as the invasion-front toads, and they no longer moved in the same direction every night.

So, with a combination of Ben's captive-bred toads and Greg's annual cohorts of wild-caught toads, we could demonstrate that evolution has occurred during the amphibian's march across Australia. Toads at the invasion front are different from toads in Queensland, and they pass those differences on to their progeny. The sedentary Cane Toads of 1935 have evolved into invasion machines. The idea that animals adapt to meet new challenges dates back to Charles Darwin, but we usually think of evolution as a slow process—tiny changes, accumulating over millennia.

How could evolutionary change happen at such lightning speed, within the lifetime of a human being? Probably by switching existing genes on and off, rather than waiting for mutation to create novel genes. It's a simple way to modify the organism without requiring major alterations to its DNA. For example, Cane Toads in eastern Australia are capable of long-distance travel, even though they rarely display that ability. If you catch a toad in northern New South Wales, drive down the road for an hour, and then release the animal, it will move much farther than usual over the next week or two, as it tries to find its way home. These displaced toads move almost as far and fast as invasion-front toads. Evolution has changed the invasion-front toad so that it expresses those abilities continuously, rather than only occasionally. The genes needed for rapid dispersal were already present in those ancestral Queensland toads—they were just diluted with a lot of other genes for stay-at-home behavior. Evolution only needed to switch on the "racehorse" genes and turn off the others.

What kinds of characteristics turn a stay-at-home Cane Toad into a racehorse? Behavior is important, but how about physical attributes as well? Racehorses are built very differently from draft horses. To clarify what was happening, I devised a miniature Toad Olympics at Middle Point in 2005. We spray-painted a circle on the ground a few meters in diameter, dropped a toad in the middle, gave it a poke, and recorded how quickly it moved out of the circle. Simple science—the sort of thing that primary-school kids might do. But those trials told us how quickly a toad could move in grass clumps compared to the open road, to check our ideas about Toad Travel Tactics. We ran the trials at night, when the toads would normally be active—and the buffalo farmers wouldn't be there to laugh at us. Our only audience was several thousand mosquitoes. As we expected, toads moved faster and more easily on the open road. But the trials threw up a result I hadn't expected. We measured the length of the back legs on each toad before running them, because some toads were longer-legged than others. When I analyzed the numbers the next day, I realized that longer-legged toads were quicker than shorter-legged toads (even at the same overall body size). With longer legs, you can hop farther. And when Greg checked his radio-tracking results, he found the same pattern: toads with longer legs dispersed more rapidly than their short-legged relatives.

Who cares if longer legs help a toad move faster? This apparently trivial result revealed an easy-to-measure trait (leg length) that is related to a toad's dispersal speed. The rate that a toad travels across country takes several days of radio tracking to measure accurately, it depends on local conditions (weather, landscape, etc.), and we can only measure it on a live toad. This makes it hard to obtain enough information to make valid comparisons. But we can measure the length of a toad's legs quickly, and even on a long-dead specimen. So now, we could test the idea that Australian Cane Toads have evolved characteristics that help them move faster and faster. Have the toads' legs gotten longer as the invasion progressed? Fortunately, Greg had been measuring leg lengths as well as body lengths ever since the first toad arrived at Fogg Dam. He didn't have a particular reason for doing it (he claims that after working on snakes for years, it was fun to measure legs), but it paid off. We could look at whether or not the long-legged toads tended to arrive first, in the vanguard of the invasion. And they did! And when Ben looked at his measurements of preserved Cane Toads in museum collections, he found the same pattern: toads from Queensland had shorter legs than toads from the Northern Territory. Thus, toads were short-legged in areas that had been colonized long ago, but long-legged at the invasion front.

We published the paper describing this rapid evolutionary change in 2006—in *Nature,* the world's premiere scientific journal. It was a bizarre story—Cane Toads have evolved longer legs, enabling them to sprint faster and faster across Australia. It was perfect material for the TV news, and for teachers to talk about in their biology classes. Our findings went ballistic in international media, from the *New York Times* to the *Shanghai Daily.* Textbooks soon included the case as an example of evolution in action. The popular media in Australia were just as enthusiastic. The *NT News,* Darwin's daily newspaper, trumpeted "Super Toad" on its front page, with a cartoon of a toad wearing a Superman suit.

The media avalanche was overwhelming. Everyone wanted to hear the story of stumpy-legged toads evolving into lithe and speedy invasion machines. That week, I did so many radio and TV interviews that I developed laryngitis. I had so many interviews on morning radio that I had to do some while sitting on the toilet. It was surreal, chatting to an audience of thousands of people while perched on the porcelain throne with my pants down around my ankles. I appeared on breakfast TV shows, with my bald noggin powdered to reduce reflections. Every taxi driver I met, as I traveled around the country, knew the story of the long-legged pioneer toads. We had truly penetrated mainstream culture, as few "science" stories ever do. It was cricket season, and I heard a radio commentator lament the inability of the English batsmen to adapt to Australian pitches—whereas, he pointed out, even Cane Toads have adapted to Australian conditions.

In 2006 we discovered another piece of the puzzle. The toads' rapid shift into overdrive is causing their bodies to buckle. That discovery came from Cathy Shilton, whose expertise in wildlife pathology is exceeded only by her tolerance in cohabiting with Greg. When she looked at preserved Cane Toads from the invasion front, Cath noticed big lumps on the toads' backbones. These proved to be spinal arthritis (spondylosis): the first report of any kind of amphibian having a "bad back." Most frogs and toads don't move around enough to stress their spines, but invasion-front Cane Toads are the exception. Spinal problems are rare in the sedentary toads from Queensland, but at the invasion front the individuals with unusually long legs—that is, the most mobile pioneer toads—are prone to arthritis. The extra pressure exerted by those long legs, and the constant travel, have taken the toads' backbones outside their comfort zone.

It's almost enough to make you feel sorry for the Cane Toad. We think of biological invasions as imposing stress on the native fauna, but we rarely empathize with the invader. Hurtling full-speed into an unknown world is seriously stressful, and Cane Toads are pushing the limits of the Amphibian Body Plan in two ways. First, they are traveling faster and farther than any frog or toad has ever done in the glorious 250-million-year history of Class Amphibia. Second, Cane Toads are among the largest frogs. Any body plan has size limits—that's why rodents don't grow as big as elephants. If you scaled its body up to elephant size, a mouse's legs would collapse under the load. Adult Cane Toads were putting their bodies under pressure even before they evolved into marathon runners; no wonder their spines crumbled as they hurtled across a continent.

Cath's discovery immediately led her to another one. Her assays revealed that those spinal bumps contained a local soil-dwelling bacterium, usually harmless to animals, that has been reported to cause problems in people with compromised immune systems. Why can't the invasion-front toads fight off these "harmless" bacteria? Sure enough, our follow-up work showed that invasion-front toads invest less in their immune systems. To move a ship at full speed, you jettison the lifeboats.

The immunological research threw up some other tantalizing results as well. For example, Greg showed that a toad's immune-system function is affected by how far it travels. Some components of the immune response are turned on by long-distance dispersal, whereas others are not. And when he tested the progeny of Ben's toads (the ones raised at Middle Point), he found a perfect match. The offspring of mobile parents had lower baseline levels of some immune responses: the ones that are activated by dispersal. A toad that is genetically programmed to move huge distances will inevitably switch System A into overdrive, and hence it evolves low baseline activity in System A. That way, it ends up with a functional immune system despite running around in ways that would otherwise knock its immune responses off-kilter.

By definition, research is a journey into the unknown. Albert Einstein once said, "If we knew what it was we were doing, it would not be called research." It's almost as difficult to obtain an unambiguous answer from nature as it is from a politician. As a result, scientists ask very small questions—like characters in a TV spoof on the trivial pursuits of unworldly academics. Unfortunately, those modest questions are the only ones we can get a firm answer to. To get at the bigger issues, we first have to solve the smaller ones, step by step. We leave the really big questions—the meaning of life, the origin of the universe, and the recipe for a perfect mango daiquiri—to philosophers and poets.

But once in a while, a question crops up that is both important and answerable. And when it does, the lucky scientist who stumbles across it goes into overdrive. I've researched a few "big issues" during my career, but the accelerating Cane Toad invasion led me into the highlight of my life in science. In 2010 we discovered a whole new mechanism of evolution.

It's actually not true to say that we (Ben, Greg, and I) "discovered" it, although I really enjoyed writing that line. Like most research, our studies built on earlier work. The process, which we named "spatial sorting," wasn't completely new. Mathematical models hinted that something peculiar was happening in range-expanding populations, but nobody had tried to fit that result into the broader canvas of evolutionary theory. We were the first to realize that something genuinely new was driving the acceleration of Cane Toads—something different from the kinds of evolutionary mechanisms described in textbooks and taught in high schools.

Ever since Darwin's masterpiece *On the Origin of Species* was published in 1859, natural selection has been viewed as the one and only mechanism that causes evolutionary change. I used to agree with that view, and I still believe that natural selection is the most important idea that any human being has ever had. There are many forms and levels of natural selection—some work on sexual traits only, and some work on family groups and not individuals—but they all rely on the same basic process. And it's absurdly simple. Evolution remains a divisive issue in some cultures, but even Special Creationists admit the reality of natural selection. It's so simple that it *must* work.

In essence, natural selection replaces traits that don't function well (that is, don't help an individual survive or reproduce) with traits that do a better job. For example, imagine a butterfly population where individuals differ in color. If some are easier to see than others, and predatory birds find and gobble down the ones that are especially obvious, natural selection will move the population average toward better-camouflaged butterflies. The mechanism is straightforward, relying on just three things: (1) more butterflies are born than will survive to reproduce, (2) better-camouflaged butterflies are more likely to survive, and (3) offspring tend to resemble their parents. If those three conditions are met, the system will evolve. Genes for camouflage will become more common, while those that make a butterfly easy to see will become less common.

Creationists accept that this kind of change occurs but argue that natural selection doesn't lead to evolution—mysterious "limits to variation" constrain the amount of change that is possible, such that organisms are stuck within "types" that can't evolve from one to the other. In response, evolutionists point to thousands of intermediate forms in the fossil record. Nobody changes his or her mind, and I don't want to get bogged down in the debate. Personally, I have no doubt that Darwin was correct: natural selection is the main reason why the Earth contains such a diverse range of creatures, all beautifully adapted to the places where they live.

But unlike most evolutionary biologists, I no longer think that natural selection is the *only* mechanism that can sculpt a new type of animal that does something more effectively than its ancestors did. In 2008, the accelerating Cane Toads convinced me that evolution can occur through space as well as through time. Natural selection is the most important evolutionary process, but now I can see a role for another process also: one that we have dubbed "spatial sorting."

The key idea behind this new hypothesis is that an invasion—like the Cane Toad's spread across the Australian tropics—is a footrace that continues across generations. Cane Toads mature and breed at about a year of age, so they have been in this particular race for eighty generations. Remember that the invasion started from the eastern coast of Australia, where the babies of those original Hawaiian toads were released by the sugar growers. Each generation, some toads dispersed farther than others. The toads that had moved the farthest—that is, those at the western edge of their expanding range—were the ones that had characteristics (like longer legs) that made them move farther. So, when it came time to breed, these more athletic toads bred with each other. They didn't have any choice about whom to mate with, because all the slow toads had been left behind. For a fast-dispersing toad at a pond near the invasion front, the only potential sexual partners were animals that dispersed just as quickly as he or she did. In our scientific papers, we cheekily called this the "Olympic Village effect"—that is, athletes having sex with each other. (The term arose from newspaper reports of the vast numbers of condoms used by athletes during the Olympic Games.)

When those athletic toads interbreed, what kind of offspring will result? If an individual's genes affect its dispersal rate, then fast-dispersing adult toads will have fast-dispersing babies. And, importantly, the mixing of genes during sexual reproduction means that some of those babies will disperse even faster than their parents. Like us, each Cane Toad gets half of its genes from its mother and half from its father. Each egg or sperm contains half of each parent's genes, sorted at random. Hence, some baby toads will (accidentally) receive genes for fast dispersal from both of their parents, thus combining those genes within a single young amphibian. For example, if the father has exceptional endurance and the mother has terrific speed, some of their babies will inherit *both* speed and endurance—and thus be more athletic than either parent. Other babies, having inherited slowpoke genes from both

During the breeding season, fast-dispersing toads in the vanguard of the invasion congregate around water bodies in the drying landscape and mate with each other, creating even faster-dispersing offspring. Matt Greenlees took this photo at a pond near Longreach in central Queensland.

of those same parents, will stay close to home. We called the process "spatial sorting" because it sorts out genes that cause faster dispersal and combines them into the same individual—thus producing a superdisperser.

Unlike natural selection, spatial sorting doesn't involve any change in the numbers of different genes through time—instead, it creates differences through space. The cornerstone of natural selection—the orthodox Darwinian mechanism—is that a gene affects its bearer's survival or reproduction. That doesn't happen with spatial sorting. In the jargon of evolutionary biology, spatial sorting doesn't require "differential fitness." It works even if being faster has no effect on your chances of surviving and reproducing. "Fast-dispersal" genes remain just as common, in relation to "slow-dispersal" genes, in the Cane Toad population as a whole. But those two types of genes end up in different places: at the front of an invasion or farther back. At the Cane Toad front line, the process constructs an amphibian that travels farther and faster than any of its ancestors. The end result is the emergence of a new kind of organism, which passes its characteristics on to its offspring. We realized that spatial sorting was an undescribed evolutionary process—the first to be recognized since 1859, when Darwin described natural selection.

The first new evolutionary mechanism since natural selection? Discovered by a few Australian bumpkins in a small village in the Australian bush, rather than by renowned professors in the halls of Harvard or Oxford? I know it sounds unlikely, but that's what happened. How can we test an idea like this? The first step is to build mathematical models, to check the logic. Ben did just that. Importantly, his models showed no benefit in survival or reproduction for fast dispersers versus slow dispersers—thus taking natural selection out of the equation. Those models confirmed our intuition: as an invasion moves forward, genes for faster dispersal inevitably accumulate at the expanding range edge. Generation by generation, our mathematical "organisms" dispersed faster and faster across the digital landscape. Even in a rigorous mathematical model in which our Darwinian Demon was sidelined, spatial sorting constructed animals that dispersed quicker than their parents. And the differences accumulated, generation by generation.

The ideas aren't relevant only to Cane Toads, or only to long legs. Because spatial sorting will operate at the edge of any expanding range, every invasion front will evolve to move faster and faster (until it hits unfavorable environments). The process will sieve out genes for animals that are more active, that have greater endurance, that forgo breeding, that invest less in immune defense, and that keep moving in the same direction every day—in short, animals like the first Cane Toads to arrive at Fogg Dam. Ever since the toads began to spread westward in 1935, any genetic change that caused a toad to slow down, or travel in a circle, or take the night off, has been left behind. And throughout that same period, genetic changes that make toads disperse faster have found their way to the front. So the Cane Toad invasion has been accelerating for eighty years because in every one of those years, interbreeding of athletic toads at the front has produced even more athletic progeny. Some of those fast-moving individuals moved back into long-colonized areas, where they bred with stay-at-home toads. But the only genes on the leading edge of the invasion were ones that made toads disperse faster. The process was slowed down a bit by the low reproductive rate of toads at the front; but when they did breed, some of their offspring were superathletic individuals that moved the invasion front forward faster and faster. It's a positive feedback loop, with every generation containing a few individuals that disperse faster than any in the generations before.

It's a simple idea, but it's difficult to wrap your head around. An analogy may help. Imagine that your body is a boat, powered by a team of ten oarsmen (your genes). Your boat is in a race with many others. Each boat has ten oarsmen, but they vary in ability: half of the oarsmen are good at rowing, the other half are not. The rowers have been allocated randomly, so on average each boat has five good oarsmen. But that's just an average: some have more, some have less. Boats with six good oarsmen and four bad ones will move quickly. Boats with four good oarsman and six bad ones will lag behind.

All of the boats start at the same time, from the same place, and race as fast as they can up the river. But after every hour, there is a ten-minute break: all the boats pull over to the bank, to the nearest picnic area, so that everyone can stretch their legs. At the end of the ten-minute break, it's back into the boats—but the oarsmen hop back into the first boat they encounter, rather than returning to the same boat as before. Just like genes in sexual reproduction, the rowers are randomly redivided among boats after every stop at a picnic area (that is, every generation). After a picnic break, no boat has exactly the same oarsmen it had before.

What will happen? After the first hour, the boats near the front of the race will be those with a higher-than-average proportion of good oarsmen. So, when the time comes for a rest break, the group in that farthest picnic area won't be 50 percent good rowers and 50 percent bad rowers anymore. Instead, good oarsmen will dominate—say, 60 percent of the people there. And when they file back into the boats randomly, this means that, by chance, there will be some boats with seven good rowers and only three bad rowers. Other boats will have the reverse mixture, and they will fall back. But the higher proportion of good rowers will allow the fastest boats to travel farther than any did in the hour before. And in that next race-front picnic break, the fast rowers are even more dominant. Say that 70 percent of the people taking a rest in the farthermost picnic area are good rowers, and only 30 percent are bad rowers. The same thing just keeps on happening: every hour (that is, every generation), oarsmen (that is, genes) that make the boat (individual) go faster will accumulate at the front of the race (invasion). And that simple process is spatial sorting. It will occur in any biological invasion, and it will make the invasion front go faster and faster.

As I've mentioned above, we didn't actually discover the process. The idea was already out there in mathematical papers. But the toad invasion made us think about it, realize its importance, and give it a name. So, in that sense, the lowly Cane Toad has affected the way that biologists think about evolution. At least some biologists. New ideas take a long time to spread.

How did we make the leap from "Here's an explanation for the accelerating toad invasion" to "Here's a new evolutionary process, different from natural selection"? It grew out of my frustration in trying to explain the idea in public talks about our toad research. To a Darwinian biologist, the obvious explanation for the accelerating invasion is natural selection—if it benefits an individual toad to be at the front line, genes for faster dispersal will increase the survival and/or reproductive output of the individuals that carry them. Thus, natural selection will produce a race of faster-dispersing toads.

That's plausible, and our data *do* show advantages for toads in the invasion vanguard. They face less competition from other toads, because they get first crack at the food resources. If that extra food helps them survive or reproduce, it will indeed result in natural selection. Hey presto—that could explain the evolution of more rapid dispersal in Cane Toads. Genes for rapid dispersal enhance an individual's access to abundant bugs. But I'm doubtful. First, toads in the invasion vanguard also run the gauntlet of predators. In later generations, the risk is lower because predators have been poisoned by eating pioneer toads. Greg's radio tracking indicates that faster-moving toads were more likely to be predated. How about reproductive output? To my surprise, toads rarely reproduced in the early days of the invasion. They just came stampeding through, heading west. Being in the invasion vanguard may reduce, not increase, a toad's survival and reproductive output.

If natural selection can't explain the accelerating rate of toad dispersal, we have to look elsewhere. Spatial sorting offers a simple alternative explanation: if you go faster, you end up in front; and that effect ramps up with every generation. Both evolutionary explanations require that an individual's genes affect its rate of dispersal, but (unlike spatial sorting) natural selection also requires an individual's dispersal rate to affect its fate. Both natural selection and spatial sorting can happen at the same time, in the same animals; accepting the reality of spatial sorting doesn't mean abandoning natural selection.

When I tried to explain this dual possibility in seminars I gave about Cane Toads in 2009 and 2010, I failed. Dismally. My audiences were confused. Natural selection was OK; people were used to the concept. But the notion of dispersal-enhancing genes hurtling westward faster than stay-at-home genes was more difficult. I agonized over how to visualize the process and what to call it. How did it differ from natural selection, and from other processes (like the delightfully named "mutation surfing") that occur at expanding range edges?

Refining those ideas was a long process. Some people think best through vigorous debates, whereas others prefer to sit back in a comfortable chair and ponder away in isolation. For me, walking is the best technique. My hero

Charles Darwin perambulated along the sandy walk behind his house in the Kentish countryside every day, pondering the mysteries of biological diversity. Like Charles, I think best when my legs are moving. This wasn't a problem on campus in Sydney—I just had to avoid colliding with students who were staring into their smartphones. But walking around lost in thought at Middle Point was a dangerous activity. Before we moved our operations to the village, Team Bufo's only base was Greg and Cathy's demountable building at the buffalo farm, where I stayed in their spare room and set up my computer on their kitchen table. In the searingly hot weather, the overhanging eaves of the building provided a shaded walkway ideal for a concept-clarifying promenade.

But this idyllic scene hid a monster—indeed, two monsters—lying in wait to pounce upon an unsuspecting intellectual. Two sociopathic bantam roosters, known as Leon and the Colonel. Despite their diminutive size—weighing less than 1 kilogram (32 ounces)—they ruled the farm. Lurking like tyrannosaurs behind the building's corner, these two miniature thugs waited until their unfortunate target took his eyes off them, then hurtled in for the kill.

Imagine yourself lost in thought, with eyes half closed as you explore some abstruse hypothesis; suddenly, two miniature demons leap up at your groin and slash out with their razor-sharp heel spurs. All thoughts of spatial sorting vanished from my brain as I shifted into fight-or-flight mode. Kicking at the psychopathic gremlins was futile; the only option was undignified retreat.

Leon and the Colonel were afraid of water, so Cathy kept a water-filled squirt bottle near the door of the demountable for self-defense. If I realized that the feathered furies were about to attack, I could repel them with a firm squirt in the face. But the pugnacious poultry were upon me lightning-fast, almost before I could get the weapon in my hand. And whenever I wasn't looking, Greg changed the nozzle setting from "jet" to "gentle mist."

The belligerent bantams set my evolutionary speculations back a month or two, but it's difficult to hold a grudge against such courageous combatants. Ironically for such a warrior, Leon eventually died of a heart attack in his sleep in 2012—probably as he dreamed of backing a bald professor in a Hawaiian shirt up against the demountable and ripping his liver out. *Vale,* Leon.

Despite the roosters' reign of terror, I made progress in understanding spatial sorting. Like most complex ideas, it made sense if you began with a simple central insight. Natural selection causes evolution through time, and spatial sorting causes evolution through space. This is heresy to most evolutionary biologists: indeed, one common operational definition of *evolution* is "a change in gene frequencies through time." But, like spatial sorting, natural selection often produces differences in gene frequencies between different areas, without causing any overall change in gene frequencies species-wide.

The most obvious effect of spatial sorting is that invasions will accelerate, but there are other effects as well. For example, spatial sorting enables invaders to disperse farther as well as faster. Any invading species eventually encounters a barrier of unsuitable habitat. If individuals in the invasion vanguard are genetically programmed to disperse as fast as possible, regardless of the consequences, then they will keep hurtling forward. Many will perish, but a few will make it through—and the species will keep expanding its range. Without spatial sorting, the invaders would never cross that barrier. During a naval battle in the U.S. Civil War in 1864, Admiral Farragut is quoted as saying "Damn the torpedoes, full speed ahead." Cane Toads at the invasion front feel the same way.

The idea of spatial sorting applies whenever a species is expanding its range—regardless of whether the invader is a Cane Toad, a Fire Ant, or a Pine Tree. And at some stage, every species has expanded its range (that's how they got to the places where they live today). In a world of changing climates, many species are changing their distributions right now. The climate suitable for a given species no longer occurs where the species occurs; the organisms have to move or perish. A species that used to live halfway up the mountain finds things getting too hot and has to move upslope. And as the species moves into that new area, spatial sorting will fashion the characteristics of individuals in the vanguard. Spatial sorting is nowhere near as important as natural selection, because it only affects genes that change dispersal rate, and it only operates in range-shifting populations. But it's big news nonetheless.

Once we got our thoughts together in 2011, we (Greg, Ben, and I) wrote a paper on spatial sorting and submitted it to one of the most high-impact scientific journals—*Proceedings of the National Academy of Sciences (USA)*. I expected a rocky reception from the reviewers, because revolutionary ideas are often unpopular. Fortunately, two of the world's best evolutionary biologists, John Endler and David Wake, saw the value of our work and supported its publication. It was my greatest career highlight to see the paper appear. And it attracted attention in the popular media as well—people were intrigued by the idea that Cane Toads have revealed a new mechanism of evolution.

Public popularity isn't the same as acceptance by professional science. Our paper didn't instantly transform the field of evolutionary biology. Most new ideas disappear into the long list of yesterday's hopes and dreams, and any new concept takes decades to win broad support. As one of its proud fathers, I've written this chapter as if the idea is important. But if you were to wander the halls of Harvard or Oxford, you wouldn't encounter evolutionary biologists squealing excitedly about spatial sorting. Most textbooks still don't mention it. Natural selection reigns supreme.

But there are encouraging signs. Mathematicians in several countries built more sophisticated models to check Ben's calculations, and all of that work unequivocally supports our idea. Indeed, the field of mathematics research now has spawned something called "the cane toad equation"—a way of exploring evolution (without natural selection) in expanding populations. It's comforting to have rigorous mathematical proof that dispersal-enhancing traits will accumulate in the invasion vanguard, even if I can't understand any of those equations. So far, nobody has come up with any contrary arguments, and the idea is appearing more and more often in scientific journals. Spatial sorting is forcing its way into mainstream biology.

We no longer have to rely on Cane Toads for empirical evidence. Starlings in South Africa have larger wings at the invasion front than in long-colonized areas. Vines in China produce lighter seeds (that disperse farther) at the expanding range edge, as do Pine Trees in North America as they spread into previously glaciated regions. In European beetles that are heading north with climate change, the winged caste is more common than the wingless caste at the invasion front. In the same area, butterflies at the leading edge of range expansion have larger flight muscles. The voles that reach far-flung Swedish islands have bigger feet than their stay-at-home cousins. These are exactly what spatial-sorting theory predicts: characteristics that enhance the rate of dispersal will accumulate at expanding range fronts.

And even more exciting, the idea is being explored in other fields. As bacteria spread out on agar plates, cells at the expanding edge exhibit mutations for faster dispersal. Bacterial cells with an extra flagellum (the little whip-like tail they use to move around) are concentrated at the outer edge of a growing colony. As in toads, pioneers with better dispersal abilities end up in front of the pack. In other laboratory studies, experimental biologists set up flour-beetle colonies and let them disperse outward—and sure enough, beetles at the invasion front evolved to move quicker and quicker, even without natural selection.

More recently, medical researchers have joined the bandwagon. A tumor that stays somewhere and grows larger isn't a huge problem, in most cases anyway. The surgeon can take it out. But if bits of that tumor evolve to disperse through the body—as Cane Toads evolved to disperse through Australia—it's a big problem indeed. And the process that drives it is spatial sorting; the best dispersers go the farthest. Understanding that process may give us powerful insights into cancer.

The key to our discovery of spatial sorting was to have the right people in the right place at the right time. Ben and I were both interested in evolution as well as ecology, so we were primed to think about how things change—how they evolve—instead of treating our study systems as static entities. As a result, our studies on a giant South American amphibian marching through Australia generated insights into the spread of tumor cells through a human body. Ain't science grand? And it's a textbook case of the value of "blue sky" research. You never know where it may lead. I'll be eternally grateful to Ben and Greg, my fellow travelers on an exciting journey that led us far beyond the ecosystems of Middle Point. Cane Toads taught us that evolution can occur through space as well as through time.

FIVE

The Impact of Cane Toads on Australian Wildlife

> The Australian outback is littered with animals such as kangaroos that died after swallowing a cane toad.
> **LIVE SCIENCE WEBSITE**

Does it matter that Cane Toads are spreading through Australia? At first sight, it's a stupid question. Hatred of Cane Toads, and the desire to eradicate them, unites Australians almost as much as a football game against New Zealand. But on the face of it, amphibians are harmless little creatures. We already have three hundred frog species in Australia. Why care about one more?

Even though Australians agree that we should eradicate Cane Toads, they disagree about why. The response in Queensland (where people have coexisted with toads for decades) differs from the answer in Western Australia (where people are bracing for the toad's onslaught). As an ecologist, I focus on the impact of invasive toads on native ecosystems. But there are other perspectives as well.

Tourism operators complain about the economic impact of Cane Toads: clients who pay for a wilderness experience feel cheated when they see feral amphibians. Otherwise, though, money isn't the big issue. Toads haven't affected any major agricultural enterprises. They eat bees around apiaries, but the impact of toads on a healthy beehive is trivial.

Cane Toads may, however, increase the abundance of bushflies—and, hence, make life less pleasant for people. Flies feed on cattle poo, and dung beetles remove that fly-food by munching their way through untold millions of cowpats every year, burying the poo for their babies to feed on. If toads eat the dung beetles, more cowpats remain—and, thus, there are more flies in our faces. One of Team Bufo's students, Edna Gonzalez-Bernal, ran some experiments to test this idea. When she added toads to an enclosure, they reduced dung-beetle numbers, which slowed down the disintegration of cowpats. And surveys in desert areas show that sites with Cane Toads have fewer dung beetles, compared to similar places that lack the voracious amphibians.

But there's a silver lining to every cloud. Increases in fly numbers may be balanced by decreases in other nuisance insects. First, toads eat a lot of bugs. Second, female mosquitoes won't lay their eggs in a pond that contains toad tadpoles—so although more flies crawl up into your nose by day, fewer mossies buzz around your head in the evening. Given the diseases that mosquitoes spread, that may be a good bargain.

For biologists, toads are inoffensive little creatures, reluctant to use their terrible poisons unless forced to do so. Most members of the public don't see it that way. The sight of a Cane Toad inspires revulsion, sometimes near hysteria. I know one intelligent, tough Territorian whose husband has to check for Cane Toads before she'll walk out the front door of her house at night. I can appreciate that phobia, because I get a creepy feeling up my spine when I encounter a large, hairy spider. I know that it bears me no malice, but it still gives me the willies.

Given that fear of Cane Toads is one of the major ways that toad invasion affects people's lives, it's interesting that nobody has tried to reduce that fear by changing attitudes. Indeed, the media hype has had the opposite effect. With tongue firmly in cheek, Greg suggested running an advertising campaign to rename the Cane Toad the "Bolivian Good Luck Frog." Tell people that seeing a toad is a lucky omen, sure to bring good fortune in love or finance—such a campaign wouldn't change the impacts of toads on native animals, but it would help people sleep at night.

But apart from the emotional well-being of our own species, the central problem caused by Cane Toads in Australia is their effect on the native fauna. A few dog-lovers would object to that summary, and Aboriginal people see the issue as more complicated (see sidebar: "Indigenous Attitudes toward Cane Toads"). But when I talk to people about the impact of Cane Toads, the fate of native wildlife is their main concern. The rest of this chapter will talk about what we know, and what we don't know, about the impact of invasive Cane Toads on the native fauna.

To put the question in context, I'll begin by outlining the effects of Cane Toads on wildlife in their native range, and in the forty or so other countries around the world to which these giant amphibians were foolishly introduced. If Cane Toads are so devastating in Australia, surely we see the same destruction everywhere else? And that simple question has a simple answer: No. There seem to be few ecological impacts of Cane Toads anywhere other than Australia.

INDIGENOUS ATTITUDES TOWARD CANE TOADS

Aboriginal people have been in Australia for so long—about 60,000 years—that it's difficult for any European to understand the relationship between indigenous Australians and the land. But the invasion of Cane Toads seems to have alarmed indigenous communities for three reasons: nutritional, social, and cultural. For groups that rely on "bush tucker," the virtual disappearance of large lizards leaves a huge gap in their daily diet. And hunting lizards is a major social activity. There isn't much else to do in remote communities, and idle hands can turn to self-destructive activities. In 2004, Kathryn Seton and John Bradley wrote that "one of the most far-reaching consequences [of toad invasion] was the stress and depression among Yanyuwa women when their daily movement across country in search of normal target prey (such as goannas and blueys) led only to 'finding Cane Toads in their holes.'" Finally, the Cane Toad's arrival poses a cultural challenge. Indigenous lore links every kind of animal to people and landscapes. A pest species, outside traditional law, doesn't fit within the existing structures.

Seton and Bradley recount some depressing scenes. For example:

> During the initial wave of cane-toad infestation this was a point of critical discussion among the Yanyuwa and Garrwa people at Borroloola. When the first wave hit the township young boys were killing the toads in their hundreds; rotting mounds of dead Cane Toads could be seen everywhere. One old man, who had as his Dreaming mother a number of native species of frogs including sand frogs and green tree frogs, called out to these boys: 'Hey leave that frog [Cane Toad] alone, it is my mother, I am guardian [*jungkayi*], you cannot kill the frog, I am crying because you are killing my mother, she has a big ceremony, you children are ignorant in these matters!' The old man's niece was sitting alongside him and she spoke back to him, 'Old man my uncle, this Cane Toad is not a frog, it has no Law, it is a stranger to this country, this Cane Toad is one that desires to turn all things into spirits [*ngabayamanthamara*], your mother would not kill her own country.' (Old Tim Rakawurlma with Annie Karrakayn, pers. comm. 1999)

Given the status of Cane Toads as Australia's "Public Enemy Number One," that conclusion is shocking. Why don't we hear similar tales of woe from other parts of the world? There are two reasons. One is that most of the ecological research on Cane Toads has been conducted in Australia. Even if the toads were mass murderers overseas as well as in Oz, we might not know about it. But the other (and more surprising) reason is that Cane Toads have indeed had little impact on native species in most other places.

I'll start with the first of these issues. Although Cane Toads have a wide native range and have been introduced to many other places, nobody cares much about them except in Australia. Latin America contains toads of many species, and to the locals the Cane Toad is just another big amphibian. The native wildlife has coevolved with toads over millions of years. Either the predators don't eat toads or they have adapted to tolerate the toad's poison.

In Australia, it's a different story. As I mentioned earlier, we have no native toads—so our frog-eating predators have not evolved to ignore toads or to resist their poisons. As a result, an Aussie predator is likely to die if it eats a toad. The other big difference is that Cane Toads are far more abundant in Australia than within their native range. In the South American rainforest it's difficult to find a Cane Toad (perhaps because there are so many other kinds of toads around), whereas in Australia it's common to see hundreds of adult Cane Toads in a single night.

But even if Cane Toads are no problem in their native range, how about in other parts of the world? The Cane Toad has a lot of stamps on its passport. Sugarcane growers and other biological-control enthusiasts introduced Cane Toads to about forty countries worldwide—from Japan to the United States, from New Guinea to Fiji. In the days before effective insecticides, bringing in Cane Toads was seen as an environmentally friendly form of pest control. Surely Cane Toads are an ecological catastrophe in those places, as well as in Australia? Well, no.

People in other countries *do* complain about Cane Toads, but never with quite the same passion as we Australians. In some areas, the reason may be a lack of research; for example, the impact of Cane Toads in Papua New Guinea probably resembles that in tropical Australia (and local tribes are still spreading them intentionally, based on the myth that the smell of toad urine drives away snakes). In Japan, we know that Cane Toads kill frog-eating snakes on Ishigaki Island, but we don't know how important this is ecologically. Recent work by Ed Narayan suggests that Cane Toads are a catastrophe for Ground Frogs in Fiji; the endangered native frogs are stressed by the presence of toads and cease reproducing.

Outside Australia, though, most of the concern comes from Florida. As in Australia, the toad's initial release generated some overblown rabble-rousing and dire warnings of danger. With scathing irony, William Reimer wrote in

1959: "Repetitiously there appear in the public press of this state startling accounts of monstrous toads which threaten housewives in their backyards, seize dogs by the head and hang on with a death-resulting grip, or attack and kill with their virulent poison the innocent neighborhood cats." However, we don't hear much about impacts of toads on the wildlife of Florida. The focus is on domestic pets, for three reasons. First, Cane Toads live in the suburbs, where the only predators they encounter are dogs and cats. Second, Florida has native toad species with poisons similar to those of Cane Toads (and, thus, native species are adapted to those poisons). Third, the ecosystems of Florida are already thoroughly trashed by other invaders. Florida has more species of alien vertebrates than it has native species. If giant Burmese pythons are spreading through the Everglades, wiping out the native mammals, you aren't going to lose much sleep over a few extra amphibians.

So, let's take a look at what actually happened when toads arrived in Australia. I will deal with the impact of the Cane Toad's parasites first, then with the invader's impact on each of the main groups of vertebrate wildlife in turn.

You never walk alone. Everywhere you go, you carry a vast number of smaller life-forms that use you as a motel room. Mites on your skin, viruses in your bloodstream, and wormlike creatures in your lungs, stomach, liver, and brain. Many of the worst impacts of biological invasions have been due not to the obvious invader (like the Cane Toad), but to the smaller life-forms that catch a free ride when their host gets moved from A to B.

The history of European empires expanding across the globe is a tragic tale of diseases spreading. Even "benign" diseases like colds and flu were fatal to native people whose ancestors had no evolutionary history of exposure. Syphilis was even worse. The success of European invaders in South America and Australia was aided by the wave of sickness in indigenous communities that came hard on the heels of the newcomers' arrival. How can you fight the invader, believing that God is on your side, when all your friends are dying of a mysterious new illness?

A similar tide of infection spreads as we move other species around the globe. For example, many species of frogs have gone extinct in the past twenty years, even in places where the habitat is pristine. We have lost some truly spectacular amphibians, such as the Golden Toad of Costa Rica and the Gastric-brooding Frog of Queensland. The iconic black-and-gold Corroboree

OTHER INVASIVE AMPHIBIANS

Cane Toads are one of the six amphibian species that are classed as major international pests. Five of the offenders are anurans (frogs or toads) and the other is a salamander. These villainous amphibians illustrate how an invasive species can spread in different ways—and cause different problems—and how apocalyptic claims of environmental doom due to the invaders are often overstated.

The first member on the Six Most Wanted Anurans List is a toad, but much smaller than the Cane Toad. An adult Black-spined Toad (an Asian species) is usually about 10 centimeters (4 inches) in length. It is small enough to stow away in ships and trucks and has been spreading rapidly across Asia (and, recently, to Madagascar). Its ecological impacts are probably similar to those of the Cane Toad, but less severe—though, like the Cane Toad, it has attracted considerable hysteria (see sidebar in chapter 7: "The INTERFET Frog").

The next villainous species is the African Clawed Frog. It's aquatic, with a flattened, egg-shaped body and webbed toes. Native to southern Africa, these frogs were transported around the world for medical use (pregnancy testing) and seem to have brought with them a fungal disease that has devastated many amphibian species.

The third unwelcome international invader is the American Bullfrog. Like the Cane Toad, it's a giant frog with toxic tadpoles. But the reasons for the Bullfrog's global diaspora are different. In some parts of Asia, Buddhists intentionally release animals into the wild for religious purposes; American Bullfrogs have been a popular choice for these ill-advised activities. More importantly, though, Bullfrogs have been moved around by people who enjoy frog-leg steaks. Bullfrogs were even brought to Australia by entrepreneurs, at the same time as Cane Toads, to set up a commercial frog-breeding company. Fortunately, the frogs perished and the exercise was abandoned. Bullfrogs would have been an ecological disaster in Australia. Unlike Cane Toads, Bullfrogs devastate the local fauna by eating it: they attack and consume large prey such as other frogs, snakes, lizards, and even birds and bats. Surveys show a consistent correlation between Bullfrog invasion and the decline of native fauna. Nonetheless, the impact of Bullfrogs isn't the only reason for that correlation. Like Cane Toads, these giant frogs prefer degraded habitats where humans have wiped out most other life. It's convenient to blame frogs, but the fault lies closer to home.

The next species in the Amphibian Axis of Evil is a very different creature: a tiny, elfin Central American frog called the "Coquí" (named after the sound of its call). Although the Coquí is only a few centimeters long, its loud call interferes with people's sleep. A chorus outside the bedroom window sounds like a truck revving its engine. As a result, house prices have tumbled as the Coquí has spread through Florida and Hawaii. And the Coquí is difficult to control, because these little frogs don't need free water to breed. A suburban garden can contain dozens of lovesick male frogs, all serenading their sweethearts. Biologists feared that the Coquí would imperil small insects, but so far things don't look too bad. The Coquí does change leaf-litter biology, but it hasn't had devastating impacts. And, with time, locals begin to regard the Coquí's call as part of the tropical night rather than an unwelcome intrusion.

The remaining troublesome amphibian is a salamander, not a frog or toad. The Tiger Salamander has a huge native range in North America, and it has been introduced to many other areas (notably in the western United States) by people using it as live bait for fishing. By hybridizing with closely related (but rare) native salamanders in California, the Tiger Salamander is eliminating the unique species that previously lived in that region. Soon, there may be no such thing as a "pure" native specimen.

The inexorable spread of these invaders illustrates the difficulty of controlling alien species. The motivations for amphibian translocations are diverse: to placate the gods, to harvest a few tasty delicacies, to catch a bigger fish, to find out if someone is pregnant. And stopping that spread is well-nigh impossible: it takes just one frog-leg enthusiast with a few Bullfrogs in his truck to found a new population.

Frog of alpine Australia is on the verge of extinction. Several threats have contributed to these declines, but one major culprit is Chytrid Fungus. The international trail of destruction by that fungus began in the 1930s, when Lancelot Hogben invented a new type of pregnancy test. If you inject the lymph nodes of a frog or toad with the urine of a pregnant woman, her hormones will stimulate the amphibian's reproductive system. Squeeze the frog a little while later, and he/she will shed sperm/eggs if (and only if) the woman's urine contains that magic hormone. And presto—you know whether or not the woman is pregnant. In 1940s Australia, such tests often used Cane Toads, so a toad's sexual products may have provided the first evidence that your grandmother was pregnant with your dad. But the most popular choice of pregnancy-tester was another type of amphibian—the African Clawed Frog—that was sent around the world as a result. Chytrid Fungus spread when these frogs escaped, or when the water from their aquarium was dumped into a stream, and it became an international plague, wiping out native frogs wherever it went.

What about Cane Toads? Did they bring alien parasites and diseases with them to Australia? To answer that question, you can dissect Australian Cane Toads and identify their parasites. Which ones were brought over with the original toads, and which are true-blue Aussies that evolved to infect native frogs and then opted for an upgrade (from the frog apartment to a roomier toad villa) when the toads came marching through? Either scenario is possible, because Reg Mungomery and his mates at the Sugar Experiment Station did a lousy job of quarantine in 1935.

The agricultural scientists who brought the Cane Toad to Australia should have kept the original Hawaiian animals isolated, bred them, and then raised the eggs in new containers. The adults should then have been killed and their bodies incinerated. If the toad importers had followed that standard quarantine practice, parasites from the adult toads couldn't have found their way into the next generation. Instead, the adult toads and their offspring all lived together in a large communal cage, where parasites in the adults could infect the young toads that were destined for release in the cane fields.

I won't talk about all of the parasite species that Cane Toads carry. First, there are just too many of them. Second, I don't know much about parasites. And third, you would have nightmares afterward and spend the next few weeks imagining that something awful is growing inside you. Instead, I'll focus on just three types of toad parasites—the ones that Team Bufo has looked at in detail.

The first case is one where Cane Toads were accused of spreading disaster but turned out to be innocent bystanders. The parasites involved are tiny protozoans called "myxosporeans," less than 2 percent of a millimeter long. They don't even have a common name. In 1986, a parasitologist named Benoit Delvinquier found a species of this group—in the genus *Myxidium*—living inside the gallbladders of Cane Toads. The same parasite occurred in some native frogs, but only within the toad's Australian range—not in other parts of Australia. Therefore, Benoit suggested that the parasite had been brought to Australia by the Cane Toad and had switched across to infect the native species. There the story sat for more than twenty years, until a Ph.D. student at Sydney University, Ashlie Hartigan, decided to check the idea. If Delvinquier was right, myxosporeans should be present only inside frogs collected after 1935 (when toads arrived in Australia). And sure enough, when Ashlie examined preserved frogs in museum collections, she found the parasite only in post-1935 specimens. Putting two and two together, Cane Toads emerged as the villain of the piece. Despite their tiny size, myxosporeans can make life very unpleasant for their hosts. It looked like the toad had inflicted some massive gallbladder aches on Australia's frogs.

Fortunately for the Cane Toad's besmirched reputation, though, Ashlie didn't stop there. She analyzed parasite DNA to check that the myxosporean inside toads was the same species as the one inside frogs. It wasn't, so the toad had been unfairly accused. Australian frogs contain at least three homegrown myxosporean species, some living inside the frogs' brains and others inside their gallbladders. Clearly, the Cane Toad didn't bring the offending myxosporean into Australia. Why did it start cropping up in Australian frogs only recently, then? Because we've trashed our waterways, and the creatures that depend on them are in trouble. A stressed animal finds it harder to fight off an infection, so that parasites like myxosporeans change from a background, low-level threat to take center stage.

The second parasite is a much larger animal (up to around 25 millimeters— an inch long—a veritable monster compared to the myxosporeans). Popularly called "tongue-worms," pentastomes have four formidable hooks around their mouth, shaped like claws, that they embed in the lining of their host's lung. The parasite's body is bloated and tongue-shaped, and a female pentastome pumps out eggs and larvae that travel through the host's digestive system to spread out and infect new insect hosts that are eaten, in turn, by either toads or lizards. And so the cycle continues. Research on pentastomes has been unpopular with scientists, owing to the ability of these parasites to transfer to new hosts. Unless you wash your hands carefully after playing with a tongue-worm, you're likely to end up with those recurved hooks slicing into your own lung lining. If you want your partner to lose weight, ask them to look up "pentastomes" in Wikipedia just before the evening meal.

The history of pentastomes in Australian Cane Toads is a convoluted tale of multiple invasions. The pentastome itself is not native to Australia, and its main host here is an invasive lizard, the Asian House Gecko. These geckos are distributed across much of the world these days, because they are good at hiding away in ships and planes. If a pale lizard flits across your hotel-room ceiling at night in Darwin or Cairns, with chirruping calls, it's probably this species. The alien lizard coexists with similar-looking native species, though, so please don't run around thumping geckos with a rolled-up newspaper.

The gecko and its pentastomes reached the Northern Territory from Asia long ago, via the narrow straits between Indonesia and the busy port of Darwin. We know it happened in the distant past, because the parasite occurs inside long-preserved geckos in the museum collection in Darwin. But there it stayed. Occasional geckos got transported from Darwin to outlying towns, inside air conditioners and so forth. But as far as we know, the pentastome was left behind. It stayed as a city-dweller. Because it couldn't develop inside native frogs or reptiles, there was no way for the parasite to reach the isolated gecko populations scattered around tropical Australia.

The Cane Toad invasion changed all that. When toads eat infected cockroaches, the pentastomes survive and develop into adults inside the toads' lungs. We know this because my student Crystal Kelehear—Team Bufo's first parasitologist—conducted research on these formidable beasts in 2011. Crystal will appear again in another parasite story below; she's a woman without fear. The pentastomes that infect Cane Toads only grow into wimpy little worms, much smaller than when they grow inside a gecko's lungs, but Crystal wouldn't have been deterred if they were the size of anacondas. I just hope she washed her hands frequently.

The tidal wave of Cane Toads that swept through Darwin in 2006 picked up the pentastomes and carried them westward (and sometimes in other directions, with east-dispersing toads on trucks). So: an invasive parasite (the pentastome) that lives inside an invasive lizard (the Asian House Gecko) has been distributed around the country by an invasive amphibian (the Cane Toad) that was brought to Australia by a fourth invader (humans). As a result of that diaspora, pentastomes will find their way into geckos in those previously isolated tropical towns. Until the toads arrived to spread the parasite, those geckos had been living a parasite-free life. I suspect that people who have been kept awake by chirruping geckos, or had one fall from the restaurant ceiling into their grilled Barramundi or glass of chilled white wine, won't be feeling much sympathy for the lizards.

The third type of parasite—another type of lungworm—played a more important role in Team Bufo's research and brought my team into conflict with other players in the Cane Toad Wars. I'll talk about those lungworms in chapter 8.

Tropical water bodies teem with fish, and any threat to recreational fishing strikes at the heart of Northern Territory culture. Will the Cane Toad invasion exterminate native fishes? Fortunately, no. Toad tadpoles feed on tiny objects, smaller than those taken by most fish, so competition for food isn't important. And toad tadpoles are inoffensive little creatures, so they won't try to eat a fish (and neither will adult toads, who couldn't manage it even if they tried). The only way that Cane Toads could affect native fish is by poisoning them: that is, if a fish decided to snack on toad eggs, tadpoles, or larger toads.

A hungry fish encountering a Cane Toad tadpole will indeed grab it, but the fish escapes poisoning because it immediately detects the toxin and spits the tadpole out. After this occurs a few times, the fish learns that these little black wriggly things taste awful, so it stops trying to eat them. The Barramundi ends up with a bad taste in its mouth, but nothing more.

Toad eggs are more dangerous. First, they are loaded with toxin, to protect them from predators. Second, that poison is hidden inside the jelly coat that is wrapped around the egg. All a predator tastes is that gooey coating, and by the time the jelly coat is digested away in the fish's stomach, it's too late. The predator gets a full dose of poison and dies.

The story of how Team Bufo discovered the danger of toad eggs is a sad one. It was 2005, and the toad invasion was just reaching Middle Point. We expected fish to be unharmed, based on Michael's earlier work, so Greg volunteered his aquarium full of pet rainbowfish and gudgeons as test subjects. All went as expected with the toad tadpoles—the fish seized them but then instantly spat them out again, with no damage to either predator or prey. But when Greg dropped in toad eggs instead of toad tadpoles, his tame fish zipped over and ate them—and didn't spit them out. Very soon we had a tankful of dead fish. That raised a thorny domestic problem: those fish were Cathy's pride and joy. Greg had a worried look in his eyes, and he chewed thoughtfully at his lower lip as he calculated the social costs of that foolhardy experiment. Fortunately, Cathy is a forgiving soul. A lesser woman would have slipped a few toad eggs into Greg's evening meal.

Although toad eggs are deadly, they only take a day or two to hatch, and they are tucked away in shallow water, where few predators will find them. The upshot is that a few fish are killed when toads spawn, but the overall impact is minor. Long-term surveys of Barramundi abundance confirm that these iconic fishes are not affected by Cane Toad arrival.

Also, fish are smarter than they look. At our field station in 2007, David Nelson ran experiments to see if a fish changes its feeding behavior after it encounters toads. He worked on Northern Spotted Gudgeons, ferocious little predators that are common in ponds around Fogg Dam. When we went out to obtain fish for David's study, the timing was perfect. A small dam had overflowed, and thousands of gudgeons swam over the dam wall hoping to find a better home—but all they found was a grassy area. As the water sank into the ground, the pasture was covered with a carpet of dying gudgeons. Most of them were consumed by egrets, but we snaffled enough for David's work.

For his experiments, David put toad tadpoles into the gudgeons' aquaria. Predictably, the fish spat them out as soon as they tasted them, and after a while they just ignored them. He then added some native tadpoles; for a while, the gudgeons ignored those as well. Eventually, though, the fish worked out which wrigglers were tasty and which were toxic, and they began to target the native tadpoles while leaving the toads alone. But, at least in the short term, a native tadpole might be safer from fish attack because of the presence of toad tadpoles.

Researchers from Macquarie University followed up this work in 2013. Their basic question was the same: "Do fish change their feeding behavior after encountering nasty-tasting toad tadpoles ?" Working on Crimson-spotted Rainbowfish in northeastern New South Wales, the Macquarie University team found that fish from toad-infested areas ignored toad tadpoles, whereas toad-naive fish grabbed toad tadpoles but then rejected them. Exactly like our results on gudgeons. But the Macquarie Uni group took it further. They predicted that in an area where toads occur, it pays for a predator to be a quick learner (to distinguish between tasty and unpalatable tadpoles). And sure enough, when given a fish IQ test, the rainbowfish from toad-infested areas learned more quickly. Ergo, toad invasion is creating a race of brainiac native fish in our ponds.

How about frogs? They live in the same places as Cane Toads and resemble them in size, shape, physiology, and behavior. They eat the same prey and are infected by the same parasites. Thus, common sense suggests that native frogs will be in deep trouble when the toads move in.

That concern mobilized frog-lovers to fight the Cane Toad invasion. The main anti-toad group that arose in the Darwin area was "Frogwatch NT," united by their fondness for these charismatic little pond-dwellers. In striking contrast to the revulsion inspired by Cane Toads, native frogs (especially green ones) evoke warm feelings in burly Territorians. Ceramic frogs adorn coffee mugs, and photographs of frogs are a common theme of posters on the wall. Any threat to our little green mates arouses great passion—and, according to the community groups, the prospects were dire. The leader of one toad-busting group painted an especially dramatic picture of life in a post-invasion world. "The swamps are silent," she claimed. "The native frogs have been wiped out."

The truth is different. As in so many other issues related to Cane Toads, intuition is an unreliable guide. The first hint that frogs were unaffected by toad invasion came from Bill Freeland, a maverick ecologist. In the 1980s, as toads swept into the Northern Territory, Bill surveyed frog numbers in advance of the toad front, and then again at the same sites a year later, after toads arrived. There was no difference. Frog numbers were unaffected at Bill's sites, and follow-up surveys in other places by other people told the same story. Peter Catling from the CSIRO asked the same question in 1999 in the Gulf Country of the Northern Territory; Gordon Grigg and his colleagues from the University of Queensland asked it across the toads' range from 2000 onward (using electronic "listening posts" to count frog calls); and John Woinarski and his team asked it with a huge survey effort across the Northern Territory that finished in 2005. Matt Greenlees asked it at Fogg Dam in 2006, and Greg Brown asked it over a fifteen-year timespan in the same place. None of these studies—not a single one—shows any dramatic effect of toads on frogs. No matter how you cut the cake, the results are clear. The numbers of frogs don't change appreciably when Cane Toads arrive.

That's crazy. It defies common sense. If you visit a billabong after toads invade, the aliens carpet the ground. But the frogs are there too. Not as big, not as obvious, and hidden away rather than out in the open (especially when toads are around). But the frogs are still there. Unlike the silent swamps visited by the community-group leader who described a post-toad Armageddon, the swamps I go to are so noisy during frog breeding activity that it's painful to the ears. And yet—common sense screams out at us that thousands of huge, invasive amphibians *must* make life more difficult for frogs. What's going on? How can native frogs flourish in a world full of Cane Toads?

As the toads approached Fogg Dam in 2004, we set out to measure their effects on frogs. Bill Freeland's data showed no impact of Cane Toads on tree frogs, but I suspected that he'd looked at the wrong species. Toads might still affect the ground-dwellers, so Matt Greenlees worked on the Giant Burrowing Frog, a toad look-alike (so much so that toad-busters often slaughter them by mistake). To run his experiments, Matt constructed field enclosures on the floodplain. Incredibly draining work in the hot humid conditions of the "buildup" season, and Matt lost a lot of weight as he dug metal sheets into the ground to form escape-proof enclosures. Fortunately, the nearby Corroboree Pub offered cold liquids that kept Matt hydrated.

As soon as the enclosures were finished, Matt set up an ambitious experiment. He put Cane Toads in some of his pens, frogs in others, both toads and frogs in yet others, and left the remaining enclosures empty. By counting the numbers of insects in each enclosure a week or two later, and recording the feeding rates of each frog and toad, Matt directly measured the impact of Cane Toads on feeding rates of native frogs. And, to my astonishment, there wasn't much effect. The Giant Burrowing Frogs didn't like their big warty neighbors and were less active if toads were around (maybe they worry about being eaten). Apart from that avoidance, though, Cane Toads didn't affect frogs. Both types of amphibians reduced bug abundance, but frogs that cohabited with toads ate just as much as those that were on their own. This perplexing result is due to night-to-night variation in weather conditions and, thus, the supply of bugs. On humid nights, there are hordes of bugs: every frog and toad can eat its fill. It doesn't matter how many competitors you have. And on nights when the air is dry, there are so few bugs that you don't catch many, regardless of whether or not a Cane Toad is sitting beside you. Competition for food occurs only on the rare nights when bug numbers are intermediate. As a result, toads usually don't compete with frogs, even though they eat the same kinds of prey.

Matt's experiment had loaded the dice in favor of detecting a negative impact of Cane Toads on frogs. With the amphibians stuck inside small enclosures, we had artificially inflated the intensity of competition. But even so, the frogs weren't affected. For the umpteenth time, I realized that despite a lifetime of studying the ecology of reptiles and amphibians, I retain an awesome ability to fail miserably when I try to predict the outcome of a simple experiment.

The impact of toads on frogs is further reduced by the toads' preference for places that have been trashed by human activities. Worldwide, the most successful invaders are species that would rather sit under a house than under a tree. Household pests like Black Rats and cockroaches (and Cane Toads) are hard to find in the bushland. Even in its native range, you're wasting your time looking for Cane Toads in the rainforest. Toads don't mind traveling through forest—usually along paths made by other animals—but they don't sit around admiring the scenery. Instead, they keep going until they find another human dwelling. Like other "weed species," Cane Toads thrive in degraded habitats.

The toad's fondness for the places we live in means that Cane Toads are less common than people think. If we see lots of them in our backyards, we assume that they are similarly common everywhere else. They're not. The highest densities I've ever seen have been around streetlights. Because toads like our company, even a few toads in the local area are enough for you to experience that squishy sensation beneath your heel as you take the garbage out at night, or discover that your boots are already occupied when you pull them on in the morning.

The toads' preference for towns also reduces their impact on frogs, because only a few native frog species are tough enough to exploit this Brave New World with its poor water quality, infectious diseases, and domestic cats. Gray-brown Roth's Tree Frogs and large, squat-bodied Green Tree Frogs gather around houselights at night in the Northern Territory, but most other native amphibians disappear when we build houses and pollute waterways. Frogs are fragile creatures, by and large, and many have gone extinct over the past century. When Cane Toads come hopping in, they target a degraded world that has already lost most of its native frogs.

Even where native frogs are abundant, they are active at different times than Cane Toads. Around Middle Point, frogs avoid the long dry season by finding a cool, moist burrow to escape the harsh conditions above ground. Keith Christian at Charles Darwin University has shown that frogs scale back their metabolism during the dry season, saving energy while waiting for the rains to return. For millions of years that was an effective tactic, but the arrival of Europeans changed the game. Now moist, food-rich patches are available all year, because we water our gardens and turn on lights at night. Cane Toads exploit that opportunity by remaining active year-round. The native frogs, locked into yesterday's plan, can't take advantage of this new situation.

The upshot is a horde of opportunistic Cane Toads, thriving year-round on resources that native frogs ignore. In ecological jargon, the invading toads have moved into a "vacant niche"—a new way of making a living, using resources that were not being exploited by the native fauna. Cane Toads can flourish in these newly created Australian ecosystems without having to push native frogs out of the way.

Nonetheless, the invasion of Cane Toads hasn't been a nonevent for frogs. For example, when Cane Toads first bred at Fogg Dam in 2006, hundreds of dead tadpoles washed up on the edges of the ponds a day or two later. They were the larvae of native frogs, fatally poisoned when they ate the toxic eggs of the toad. But although catastrophic for those individual tadpoles, the mass poisoning had little effect on overall frog numbers. Despite the carnage, thousands of young frogs emerged from those ponds. Indeed, a massacre like this can *increase* the numbers that survive to metamorphosis, by reducing competition for food. A few hundred deaths make life easier for the survivors.

What happens if the encounter between native and alien amphibians occurs a few weeks later, after tadpole life has finished? A few unlucky frogs are eaten by toads, and a few frogs are poisoned when they try to eat baby toads. But neither interaction happens often enough to affect the overall numbers of frogs.

In summary, toads sometimes poison frogs, compete with them, and eat them—but the impacts are minor compared to climatic factors. Heavy rain at the right time can fill the breeding ponds and result in a million baby frogs, whereas dry weather can eliminate those ponds and kill a million tadpoles. In this dynamic system, any effect of Cane Toads on native frogs is trivial compared to the impact of rainfall variation or to the ecological changes wrought by humans.

Moving on to the reptiles, I'll start with turtles (or "tortoises," as many Australians call them). One of the great sights of Fogg Dam is the annual turtle migration. Thousands of Long-necked Turtles climb out of the dam and scramble across the dam wall as soon as wet-season rains fill the floodplain below. And when the floodplain dries out a few months later, the turtles trudge back the other way, to wait out another dry season in the dam. It's a perilous journey. Sea Eagles swoop down by day, and Saltwater Crocodiles lie in wait at night. To a big croc, a turtle is like a chocolate-coated ice-cream treat—crunchy on the outside, juicy on the inside. Big crocs line up downstream of the dam wall, mouths open in the torrent. And the next morning, perforated turtle shells wash up at the water's edge. I shiver when I think of the power involved—the idea of jaws that can punch a hole through a turtle shell is, frankly, bowel-loosening.

Turtles are targeted by other predators as well. On many nights that Greg and I walk along the dam wall looking for snakes, we encounter Aboriginal people on the lookout for a traditional "bush tucker" meal. But despite this seasonal slaughter, Long-necked Turtles are common in Fogg Dam. The presence of people may even protect turtles from more formidable enemies. In some parts of the Northern Territory, turtle numbers have plummeted as a result of predation by feral pigs. Rooting through the dried billabongs every year, pigs kill and eat every turtle they can find.

With turtle populations under pressure, what is the impact of Cane Toads? Probably not much. Many native turtles are herbivores, so they aren't poisoned by Cane Toads. The giant Pignose Turtle fancies waterweeds, and the Short-necked Turtles like vegetable material (although they've been seen gobbling up toad tadpoles without ill effect). Some carnivorous turtles tolerate the toad's poisons and are well-nigh invulnerable. Tim Hamley and Arthur Georges kept two Snapping Turtles in captivity for four months, feeding them only on Cane Toads!

The only shell-bound victims of the toad, so far as I know, are the longnecks. Ecologically, a Long-necked Turtle is a snake inside a shell—an ambush-hunter that strikes out at passing prey. The responses (and fates) of long-necks are similar to those of fish. Toad tadpoles are seized enthusiastically, then spat out. The turtle escapes with a bad taste in its mouth. Toad eggs are more dangerous because the poisons are hidden beneath a nontoxic jelly coat. When that covering dissolves inside the turtle's stomach, the sudden rush of poisons is fatal.

Crocodiles were rare—or, at least, rarely seen—when I began working in the Northern Territory thirty-odd years ago. In the 1980s, seeing a big croc was a thrill. During many months of fieldwork in Kakadu National Park, only one really large Saltwater Croc turned up on a regular basis. Five meters (16 feet) long and as fat as a Water Buffalo, "Albert" was less wary than most of his relatives; he enjoyed lying up on the bank rather than hiding in the murky water.

Times have changed. Now, crocs are abundant. Professional hunting was banned in the 1970s, but recovery was slow until a new, evidence-based approach to management was adopted. Spearheaded by Grahame Webb, it is one of the most successful conservation programs ever implemented worldwide: a rare example of a large predator brought back from the brink of

DIFFERENT, BUT NOT INFERIOR

For a vertebrate animal, there are two ways of making a living: endothermy ("warm-bloodedness") or ectothermy ("cold-bloodedness"). Mammals and birds are endotherms, relying on our high metabolic rate to keep ourselves warm. That evolutionary strategy is expensive: more than 90 percent of all the energy we consume in our food is burned up to keep our bodies at a constant high temperature. The cost is offset by a high benefit, though: we can run fast and think quickly. We're like race cars: high fuel consumption, high performance.

The other strategy is ectothermy, as seen in reptiles and amphibians. A Cane Toad doesn't need to eat to keep warm—instead, it functions at environmental temperatures and looks for a hotter location if it needs to warm up. Ectothermy is cheap, but there's a downside as well: a cold toad is a slow toad. Ectotherms are like bicycles—cheap to run but (relatively) low performance.

Neither of these strategies is "better" than the other. If fuel is cheap and always available, the race car has an advantage. In the natural world, that would mean ecosystems with fertile soil and good rainfall, which ensure abundant food. And a race car is especially useful if there are a lot of steep hills (periods of cold weather) that require energy to surmount. You can travel wherever you like, whenever you like, and you can afford to pay the bills.

But if fuel is expensive, and the world is flat, endothermy is costly without conferring much benefit. If soils are poor and rainfall is erratic, food is difficult to find. If the rain stops for months at a time, you're in trouble. Your fancy race car is useless if you can't buy fuel for it. And, in hot weather, it's easy to stay warm without spending energy—the bicycle comes into its own.

In tropical Australia, bicycles (which we call "pushbikes") reign supreme. Australia is very old geologically, without glaciers or volcanic activity to bring new soil to the surface. Soil nutrients have leached down over the millennia, so they are no longer available to plants. Rainfall is unpredictable and sparse. Food is hard to find. And, lastly, the wet–dry tropics are hot year-round. The maximum air temperature in Darwin is above 30°C (86°F) almost every day. Why waste 90 percent of your energy on keeping warm? Ectotherms win hands-down when it comes to energy efficiency.

This is a radical argument. We've been brainwashed about "higher" and "lower" animals ever since we were in primary school, and we always imagine our own species comfortably perched on the

> top of the evolutionary tree. It's difficult to believe that a Cane Toad is just as sophisticated as we are. But, in the right circumstances, "primitive" ectothermy is a better tactic than "advanced" endothermy. For example, consider the mighty crocodile. The structure of its heart, lungs, and bones indicates that its ancestors were endotherms, but a few million years ago it decided the rent was too high and it would downsize to a cheaper lifestyle. If you have to lie around for hours in warm water waiting for a wallaby to come down to drink, ectothermy enables you to wait patiently until the next marsupial-flavored snack hops into view.

extinction. Grahame's key discovery was that in nature, most crocodile eggs are drowned by flooding or are eaten by goannas. That provides a simple way for managers to increase croc numbers: collect eggs from nests just after they are laid, incubate them in the lab, and release the hatchlings back into the river, thereby eliminating the high egg mortality. You can even retain most of the hatchlings, and raise them for the commercial skin industry, but still release enough to build up the number of baby crocodiles in the wild. The profits from their captive siblings can sustainably fund the enterprise. Of course, collecting the eggs is not an easy task; Grahame bears an impressive bite mark on his leg from a nest-defending mother croc. But his plan worked. So if you visit Darwin today, you'll find farms that incubate eggs, raise hatchling crocodiles, and make money out of them. A bounty from egg collecting motivates landowners to preserve habitat and big crocs rather than (illegally) shoot them. And, last but not least, we have one of the few commercially successful industries in tropical Australia.

Given the current abundance of Saltwater Crocodiles, it's fortunate that they usually ignore the dining opportunities offered by drunken locals and naive tourists. Nonetheless, the Darwin newspaper runs frequent front-page stories about "crocodile attack," usually when an inebriated idiot gets nipped while harassing an innocent reptile.

Tropical Australia has two species of crocodile—the giant Saltwater Croc (or "saltie") and the smaller Freshwater Croc (or "freshie"). Salties attract most of the media attention, especially when they eat someone. Salties aren't at risk from Cane Toads, for two reasons. First, they are physiologically resistant to the poison, perhaps because their range extends into Asia, where they

Freshwater Crocodiles are killed by eating Cane Toads in some areas but are unaffected in other regions. We really don't understand why. Photo by Greg Brown.

have evolved side by side with native toads. And second, salties are big. It would take a massive dose of poison to kill a 1,000-kilogram (2,200-pound) crocodile. Saltie populations don't decline when Cane Toads arrive. We know this because crocodile numbers are counted accurately every year in the Northern Territory. It's another advantage of "sustainable use": if animals are worth money, managers keep a close eye on them.

The Freshwater Croc is a very different animal. Abundant in the upper reaches of many tropical rivers (away from their voracious larger relatives), freshies often eat native frogs—and thus are inclined to chomp down on a Cane Toad. In a few river systems, that behavior has had disastrous consequences. Up to 90 percent of freshies are fatally poisoned as soon as Cane Toads arrive, and the rivers are littered with the floating carcasses of dead crocs. Remarkably, though, the freshies at Fogg Dam were unaffected: they ignored toads rather than trying to eat them. I often see a freshie hauled out on the dam wall, studiously ignoring the Cane Toad beside it.

What's going on? Why did freshies die on the Victoria River, but not at Fogg Dam? One of my Ph.D. students, Ruchira Somaweera, tried to answer that question, but the more we found out, the more confusing it became. First, Ruchira's experiments showed that young freshies are smart. They learn not to eat Cane Toads after only a single bad experience (mouthing or eating a small toad). So the babies aren't at risk. But why are adult crocs killed by toads in some places but not others? We still don't know, but a possible answer was suggested to me recently by Dave Lindner, an ex-ranger from Kakadu. Dave's idea is that a predator that seizes an adult toad in the water will survive, because the water washes away the toxin—just as a vet will do when a dog is poisoned by seizing a toad. But if the freshie crawls up on land to get its meal, as may happen more often around the smaller streams, it receives the full dose and dies before it can return to the water.

The savanna woodlands of the Northern Territory abound in terrestrial reptiles, including spectacular giants like goannas, pythons, Frillneck Lizards, and King Brown Snakes. How did the Cane Toad invasion affect tropical reptiles? The news isn't good, although there are silver linings to the cloud. For example, the toads don't compete with reptiles as much as you might expect. Cane Toads eat large numbers of ants, whereas most lizards avoid them. Similarly, direct predation by toads isn't important either. Cane Toads consume a few small snakes and lizards—including the elegant Burton's Legless Lizard, one of my personal favorites—but not enough to affect overall numbers. The only reptiles that are in trouble are the species that eat frogs, which will try to snaffle up a tasty-looking Cane Toad when the invaders first arrive.

We began to find dead goannas as soon as toads reached Fogg Dam in 2005. The main victim, the Yellow-spotted Goanna, is one of the world's most spectacular lizards. An adult male can grow to more than 2 meters (6 feet) in length and weigh more than 7 kilograms (15 pounds). That's a lot of lizard. In places like Africa and Asia, the top predators are lions, tigers, wolves, and so forth. But those warm-blooded animals need more food than the harsh, stochastic Australian environment can provide (see sidebar: "Different, but Not Inferior"). As a result, goannas fill the ecological roles taken by large mammals on other continents. Before toads arrived at Fogg Dam, I often saw several goannas during a single day's fieldwork. Fantastic creatures: stalking imperiously across the floodplain, tongue-flicking, with more than a glimmer of intelligence in those big, alert eyes.

Although these giant lizards are called "goannas" all over Australia, many scientists screw up their prissy little noses in horror at the term. "It's a mistake," they sneer. "The word *goanna* is a corruption of *iguana*." Iguanas are found in other parts of the world, including the Cane Toad's native home. They are almost as big as goannas, but the resemblance ends there. The American lizards are sleepy vegetarians, spending their time draped across a branch high in the forest and chomping down on an occasional leaf. Judging by the look in their eyes (in the brief intervals when those eyes are open), those leaves contain a soporific drug. In contrast, goannas are more like miniature velociraptors.

It's a shame we named our killer reptiles after wimps from South America, but I like the word "goanna": it captures the rough affection that Australians feel for this extraordinary beast. Goannas are voracious hunters, adept at finding small creatures hidden in the grass. And, unfortunately, Cane Toads are easy to find; unlike native frogs, the toxic foreign amphibians don't bother to hide. The first toads to reach the Fogg Dam floodplain were soon discovered and seized by hungry goannas—with devastating results. I found one dead goanna with a live Cane Toad beside it, almost unharmed. A single Cane Toad might kill two or three goannas before being terminated by one of them.

We were studying goannas at Fogg Dam from long before the toads' arrival, so we knew how many goannas had lived on the floodplain and how many survived. The carnage was awful; 95 percent of our goannas were killed in the first few months of 2005. It was gut-wrenching. I still see goannas at Fogg Dam, but they are a tiny remnant of a once dominant monarchy.

The mass mortality of goannas at Fogg Dam echoed similar catastrophes that had been noted (anecdotally) for the same species earlier in the Cane Toad invasion. When 95 percent of the goannas are killed, people notice it. But the second lizard species annihilated by toads was a real shock; nobody had realized that it was at risk. The Northern Bluetongue Lizard, a colorful relative of the iconic Bluetongue Lizard of southern Australia, is a heavy-bodied, short-legged lizard that can reach half a meter (1.5 feet) in length. Blueys were abundant around Fogg Dam in the decades before toads arrived, and we commonly saw them crossing the road during both day and night. Blueys have very catholic diets, eating plant material as well as any small animal slow enough to be caught. So I listed this subspecies as "possibly vulnerable," but I thought they were on the safe side of that list. Our rates of encountering blueys varied through the years, and I wasn't worried when we didn't see these reptilian roustabouts for a few months after the toads invaded. But as time dragged on, it began to look as if something sinister had occurred. A check of Greg's field-survey notes in 2008 confirmed that he hadn't seen any blueys in the three years since Cane Toads had appeared. Not a single one.

Samantha ("Sam") Price-Rees was just beginning her Ph.D. at the time, so she took on the Puzzle of the Disappearing Lizards. She soon confirmed our fears. Sam's laboratory experiments showed that blueys readily attack Cane Toads, with fatal results. A mouthful of toad poison is the end of the line for one of these giant reptiles. Sam couldn't radio-track blueys around Fogg Dam, because there were none left, so she worked instead at the edge of the toad invasion near the Western Australian border. Sure enough, the arrival of toads was followed by a wave of lizard mortality. It was heartbreaking to follow the telemetry signal through the bush and find rotting remains instead of the bright-eyed, vigorous reptile that you'd radio-located the day before.

The problem isn't restricted to the tropics. Cane Toads are also wiping out lizards in southern Australia. In 2013, my student Chris Jolly surveyed wildlife at picnic grounds along the New South Wales coast and found a huge decline in large lizards (Water Dragons, Land Mullets, and Lace Monitors) after toads arrived. Many more people live along the New South Wales coast than up in the tropics, so why didn't anyone notice this carnage? It's a sobering example of how people can fail to notice an ecological catastrophe even when it is happening in their own backyard.

Snakes are my passion, and they were the animals that first attracted me to tropical Australia. For someone who likes serpents, this is heaven. The billabongs teem with Arafura Filesnakes—strange, nonvenomous fish-eaters with saggy-baggy skin. Other harmless snakes slither through the trees and on the ground, and swim in the mangrove creeks. On land, spectacular pythons—some of the largest snakes in the world—are extraordinarily abundant. And the venomous species that give Australia so much notoriety, and scare the bejesus out of so many tourists, hurtle across the road in front of you as you drive along.

We expected the local snakes to go into sharp decline as soon as toads arrived. From studies that Ben Phillips had done during his Ph.D. work, we knew that most Australian snakes can't survive a dose of toad poison. A snake that ate a big toad was dead within minutes. Ben stayed on with Team Bufo after his Ph.D. was completed, and he was based at Middle Point in 2005 when the toads came flooding in. He decided to find out—in detail—what was happening to our beloved serpents.

The species that Ben chose to work on was the floodplain Death Adder, a short, heavyset snake that ambushes its prey rather than actively searching for it. Ben collected adders from the Fogg Dam floodplain before the toads arrived and kept them captive for a year. Then, as soon as the toads arrived, Ben fitted his captive snakes with tiny radio transmitters and released them back onto the alien-infested floodplain. Daily checks of the telemetry signal confirmed our fears. A massacre was in progress. Half of the adders were dead within a few weeks, usually found with toad corpses lying beside them. Most of Ben's snakes hadn't been able to resist the deadly new amphibian. So, for the first time, we didn't have to rely on anecdotes—we had the smoking gun. Cane Toads slaughter Death Adders, in the wild as well as in the laboratory.

For other species of snakes, though, the evidence of toad impact is less compelling. That may say more about the difficulty of accurately assessing snake numbers than about the impact of Cane Toads. Anyone who walks the same route night after night—as Greg does, on the Fogg Dam wall—encounters massively variable snake abundances. A dozen snakes one night, none the next. Snake enthusiasts attribute that variation to weather conditions, but when Greg and I analyzed his data in detail, we discovered that the story is far more complicated. The number of snakes that Greg found on a given night wasn't consistently related to temperature, rainfall, wind, cloud cover, moon phase, humidity, or anything else we could think of. Snake counts often differed between roads a few kilometers apart—a good night for snakes on the dam wall was often a bad night on the nearby Arnhem Highway, and vice versa. It's difficult to measure the true abundance of uncommon species. That's a shame, because one of those species—the King Brown Snake—is a truly magnificent reptile, and I would dearly love to know more about it.

King Brown Snakes are among the largest venomous snakes in the world. Although I never conducted any detailed research on them at Fogg Dam, 3-meter-long (10-foot) kingies were a highlight of my fieldwork there in the 1980s and '90s. When Thomas and I set out at night to collect Water Pythons, there was always a chance of coming across an awe-inspiring kingie as well. Fortunately, big King Brown Snakes are mellow, probably because nothing is silly enough to attack them. We wouldn't see any kingies for weeks at a time, and then—for no apparent reason—would find two or three in a single night. But, sadly, these giant snakes have virtually disappeared from Fogg Dam. Was this due to Cane Toads? There's no doubt that kingies eat toads and die as a result. But Greg's nightly counts showed that kingie numbers were declining by 2002, three years before the toads reached our study site. A more

Accustomed to eating native frogs, King Brown Snakes often seize Cane Toads and die as a result, because they are very vulnerable to the toad's poisons. Photo by Jonathan Webb.

important factor was probably the collapse of the banana industry, which resulted in dense, rat-rich plantations being transformed into open mango orchards. The toad invasion is only part of the reason why King Brown Snakes have disappeared from the floodplains of tropical Australia.

Are birds at risk from the toad invasion? Most birds are too big to be eaten by Cane Toads, but once in a while an unusually large toad eats an unusually small bird. As the toad first spread through Queensland in the 1930s, a Brisbane newspaper published the following letter from a reader:

> We had a hen that used to go under the kitchen floor to lay, where it couldn't be reached, and on one occasion it came out proudly leading no fewer than 23 chicks. We noticed that one of these disappeared every day, and after a week, when half a dozen had gone, we heard squeaks under the kitchen one night and went with a light to investigate. We were greatly surprised to find a large toad squatting close to the hen, and we actually saw it drawing out one of the

chicks from under her. It had the chick gripped by the crop. When we poked it with a stick it dropped its prey, which at once scurried back under its mother.

Toads are clumsy climbers, so a bird's nest up in a tree is safe—but a chick inside an underground burrow may be in peril. Two studies have looked at Rainbow Bee-eaters, gorgeous fast-moving birds that flash metallic colors in the sunlight. Bee-eaters nest in burrows dug into the ground, rendering them vulnerable to a marauding toad. In a study conducted in Queensland in 2004, Chris Boland reported that Cane Toads enter the burrows of bee-eaters and devour their eggs and chicks. But follow-up work by Christa Beckmann at Fogg Dam in 2009 revealed a more encouraging story. Christa found toads in only a few of the local bee-eater burrows—those close to the water's edge. So even if toads eat baby birds at Fogg Dam (and Christa didn't record this), few nests are affected.

The reverse interaction—bird eats toad, rather than toad eats bird—might be more of a concern. Predatory birds like herons, egrets, hawks, and kookaburras all dine on native amphibians. If those birds also eat Cane Toads, and can't tolerate the toad's poisons, the invader's impact on avian predators might be as catastrophic as it has been on reptiles.

Toad-busting groups have claimed that bird populations are devastated, but I'm skeptical that such a disaster would have escaped public attention. Australia is full of obsessive bird-watchers who can distinguish between a sparrow and a starling from the other side of the paddock in heavy rain. And bird enthusiasts publish prolific reports about their feathered subjects in specialist journals. Where are the reports of Death By Toad? Vulnerable predators that eat Cane Toads suffer a massive heart attack and drop dead almost immediately. Birds are easy to observe, and every Australian school-kid has seen a Magpie grabbing worms on the back lawn. If Cane Toads fatally poisoned birds, there should be eyewitnesses—followed by screams of outrage, heartrending photographs of dying birds, and so forth. Little old ladies in sensible shoes marching toward Parliament House with hatred in their hearts, vengeance in their eyes, and a dead parrot named "Paul" cradled in their arms. And that hasn't happened. Likewise, many otherwise sane people regularly get up before dawn to go and stand out in the drizzle and count birds. As a result, numbers are censused more often for birds than for any other wildlife group. And there is no evidence that any birds have become less common since Cane Toads arrived. Not even a hint.

Christa and I reviewed everything that had been published about interactions between Cane Toads and Australian birds. A bird enthusiast herself, Christa enjoyed trolling through amateur bird-lover magazines as well as professional journals. We concluded that although 180 of Australia's bird species overlap with Cane Toads geographically, there are only a handful of records of birds eating toads, and even fewer of any negative effects. And the stories are inconsistent. A few kookaburras and crows have died after eating toads, yet others (of the same species) have shown no ill effects.

To find out why Aussie birds haven't died in droves, Christa conducted field experiments at Fogg Dam in 2009. She placed shallow, water-filled trays containing Cane Toad tadpoles (or juveniles) on the dam wall, among hundreds of egrets that were scarfing up small fish and native tadpoles. The wading birds picked out fish that Christa put in those trays, but they left the toads alone. Australian birds know full well that Cane Toads are a poor dietary choice.

At Fogg Dam, the only birds that regularly eat Cane Toads are hawks (Whistling Kites and Black Kites) that cruise the roads at dawn looking for roadkill. Anything that has been run over (from wallabies to grasshoppers) is fair game. To look at this scavenging in detail, Christa carefully laid the bodies of road-killed frogs and toads in trays beside the road. The kites took the frogs but rejected the Cane Toads. In the dry season, when food was scarce, the birds also took a few toads—but they only ate the tongue, leaving behind the poisonous parts of the toad's body. And if they had a choice, the kites ignored large toads in favor of smaller (less poisonous) juveniles.

Evolutionary history explains why Cane Toads pose little risk to Australian birds. Birds can fly, creating genetic exchange between populations in tropical Australia and those in Southeast Asia. Indonesia is just a hop, step, and jump away from Darwin, and there are plenty of native toads in Asia. They aren't as big or as toxic as Cane Toads, but they produce similar poisons. The ancestors of Australia's tropical birds evolved the ability to detect the toxin. Kites will readily eat a hamburger pattie left on the road, but they will reject one that has been smeared with Cane Toad poison. Judging from the expressions on the hawks' faces, toad toxins taste awful. An international exchange program has paid off for birds, and the lack of such a program has been devastating for snakes and lizards.

What are Australia's most iconic animals? I'd vote for a Frillneck Lizard or Red-bellied Black Snake, but I'm in the minority. Most people prefer furry fluff-balls to scaly serpents, and a popular vote would hand the crown to some kind of marsupial. In terms of cuteness, it's hard to go past the quoll—a marsupial carnivore the size of a domestic cat, with adorable big brown eyes, an ever-twitching nose, a bushy tail, and a severe case of attention deficit disorder. Quolls are constantly on the go, at a feverish pace. And they have bizarre love lives—they grow fast and die young. Male quolls hurl themselves into their first (and only) mating season so passionately that their bodies wear out. For a male quoll, sex is terminal; the females are left to rear their babies alone. In their quiet moments after the males have self-destructed, female quolls must sit around reflecting that they have just had one hell of a mating season.

Quolls are in desperate trouble, like all their relatives. The carnivorous marsupials (Family Dasyuridae) have fared poorly since Europeans arrived in Australia two hundred years ago. The largest dasyurid was the Thylacine (or Tasmanian Tiger), exterminated by guns, dogs, and land clearing at about the same time as Cane Toads were introduced to Queensland. The second-largest dasyurid, the Tasmanian Devil, remained abundant until the 1990s, when an infectious tumor disease appeared in eastern Tasmania. Devils are pugnacious brutes, often fighting each other for access to carcasses. The tumor, spread when one animal bites another, is invariably fatal. Thus, the Tasmanian Devil is following its larger relative into steep decline and, perhaps, extinction. Continuing the theme of "different threatening processes but the same dismal result," the next-largest dasyurid, the Spotted-tail Quoll, has been poleaxed by urban expansion in mainland Australia. Once widespread, it now clings precariously to existence in a few pockets of forest. And the smaller Northern Quoll, common throughout the Australian tropics a century ago, has vanished from most of its former range. Isolated populations remain in scattered rocky hills.

It's this last species—the Northern Quoll—that has been unlucky enough to meet the Cane Toad. The fast lifestyle of quolls gives them an enormous appetite, and they aren't choosy about what they eat. A hungry quoll that encounters a Cane Toad is likely to make a fatal mistake. The arrival of toads spells disaster for quoll populations—although, to be fair, this is just the final nail in the coffin. Populations of Northern Quolls were declining long before Cane Toads arrived. We don't know why—perhaps the culprit was changing fire regimes, or feral cats. Regardless, the Northern Quolls were only a shadow of their former might by the time the toad arrived.

So, what can we conclude about the ecological impact of Cane Toads in Australia? It has ranged from catastrophic to trivial, depending on what species we are discussing, and even which population of that species. Nonetheless, strong patterns have emerged from our work. The toad's greatest impact is via fatal poisoning. In many native species, occasional individuals have died after eating Cane Toads. In a small number of species, the impact has been far worse: populations of goannas, Bluetongue Lizards, Freshwater Crocodiles, and Northern Quolls are well-nigh eliminated within a year or two of toads arriving. That casualty list is awful, but it's only a tiny fraction of tropical wildlife diversity; the impact wasn't as cataclysmic as I had feared. Thus, the "big picture" is of a disaster for a few species (the large predators) whereas everybody else dodged the warty bullet. It took another year or two for us to discover how they did so, and that is the subject of the next chapter.

SIX

How the Ecosystem Has Fought Back

> Sweet are the uses of adversity,
> Which, like the toad, ugly and venomous,
> Wears yet a precious jewel in his head.
>
> **WILLIAM SHAKESPEARE**, *As You Like It*

So far, my conclusions about toad impact may not sound as horrendous as you expected. For decades, a huge media blitz has proclaimed that Cane Toads are an across-the-board ecological catastrophe—that the toad invasion wipes out all living things, like a toxic glacier scraping everything before it.

When Cane Toads first arrived at Fogg Dam, I expected widespread ecological carnage. What the toads didn't poison, they would eat. When that didn't happen—when it became clear that the toad's impact was falling on a few species of large predators, but nobody else—I feared that perhaps it was just too soon to tell. But as the years went on, and our datasets accumulated, I began to relax. By 2008, it was clear that the toads were not having as severe an impact as I had feared.

In retrospect, my main error had been to assume that native wildlife would be helpless in the face of the Great Toad Invasion. I feared a repeat of the catastrophe brought about by earlier biological incursions, such as the marches of the Rabbit, Fox, and domestic cat through Australia, or that of the Kiore (Pacific Rat) through New Zealand. But as we measured the impact of toads on wildlife, we learned that native animals were far from helpless. They were adapting to the presence of toads using two different mechanisms that worked over different timescales.

The first of these mechanisms happens quickly and is familiar to all of us. If you eat something that makes you ill, you don't want to eat it again because the taste and/or smell of that substance nauseates you. The process is called "conditioned taste aversion," and I can personally testify to its power and longevity. On a camping trip when I was seventeen years old, I became far too closely acquainted with a large bottle of Scotch whisky. I was deathly ill for most of the next day, and the aroma of Scotch still makes me nauseous.

Any species with a broad diet can benefit from the ability to learn what kinds of foods are good to eat, and which ones should be avoided. If a type of food makes you feel ill, don't eat it again. That is what happens when most native predators encounter Cane Toads. They seize the first one with glee, but—even before they have gotten the first mouthful down into their stomach—they realize that something is wrong. Very wrong. Toad poison tastes terrible. Most animals that grab a toad spit it out and then regret their foolishness. In the 1970s, before the advent of safety committees in universities, biologist Richard Wassersug asked his fellow students to hold tadpoles in their mouths to see how they tasted. The volunteers all agreed that the Cane Toad tadpole tasted revolting.

Conditioned taste aversion is easy to study in the lab. First, obtain some toad-naive predators—such as fishes, frogs, lizards, or marsupials—from areas that haven't yet been invaded by toads. Second, offer them a small toad and watch what happens. So long as you use small toads, a predator that eats a toad won't be killed—but it will become ill. After it recovers, offer it another toad. And again watch what happens.

The critical insight came when Jonno Webb and I ran trials on Planigales at Middle Point in 2007. These small brown marsupials look like mice, but appearances are deceptive: they are ferocious, shrew-like predators. Planigales are delightfully cute, with large dark eyes, twitchy whiskers, and lightning-fast movements. They are pocket-sized relatives of the quolls that had been devastated by toad invasion, and we feared that the same fate would befall these small, hyperactive marsupials. The toad tsunami had just hit Middle Point, and we were frantically testing species after species, trying to find out which ones were in peril. Planigales were an obvious priority, although the prospect of poisoning them in a lab trial was awful. After trapping twenty Planigales at Fogg Dam, and getting them settled into captivity and feeding happily on crickets, we couldn't delay any longer. One evening, Jonno and I dropped small toads into the Planigales' cages, hoping that they would ignore these toxic morsels.

But they didn't. Every Planigale raced out of its shelter, grabbed its toad, and started chowing down. Within minutes, every cage contained a quivering little furry body, lying on its side and close to death. It was horrible. And worse, it meant that toad invasion would kill hundreds of thousands of

Planigales across the tropics. Jonno and I were grim-faced as we walked around scoring "not moving, probably dead" in cage after cage. But then, ten minutes later, a miracle occurred. A Lazarus moment. One after another, the Planigales came back to life. We had chosen very small toads, hoping that they wouldn't contain enough poison to make a deadly meal. And it worked. Three of the little carnivores never woke up, but the rest were soon bounding around their cages again—although looking a little worse for wear.

Had they learned their lesson, or would they grab another toad and, therefore, eventually end up with a fatal dose of toad toxin? The only way to find out was to throw in another toad. And in a completely unprofessional manner, Jonno and I both squealed with delight when, one after another, the Planigales approached the new toads cautiously instead of at full speed, sniffed the interlopers carefully, then zipped back to their shelters without launching an attack. Crickets were gobbled up, but toads were ignored. Jonno and I went to our beds that night after a few glasses of wine, with smiles that we couldn't wipe off our faces. A few Planigales would die as the toad front rolled through, but the population would survive.

And Planigales weren't the only genius animals. To my astonishment, many other native species exhibited equally rapid learning. I had assumed that these "simple" creatures were inflexible little robots, and I was delighted to be proved wrong. I should have realized that predators often encounter noxious insects and benefit by learning to leave them alone. The response to toads just built on that flexibility. The A-grade students included small-brained creatures like frogs and fishes as well as marsupials. For all those species, toads would have only a minor impact. That's not much comfort if you're the unlucky frog that eats a toad big enough to kill you, but it's great news for the ecosystem. For most native species, the Cane Toad invasion is not a death sentence.

If native predators are so smart, why do some of them—like goannas, Bluetongue Lizards, Freshwater Crocodiles, and quolls—die instead of learning? Because they are large enough to attack adult Cane Toads. The outcome of an encounter between a predator and a toad depends on the size of the toad. A small toad has very little poison, whereas an adult toad is chock-full of it. Even gram for gram, a small toad is much less deadly than an adult. Small predators attack toads that contain only enough poison to make them sick, not to kill them. Before long, the predator learns that it's a bad idea to eat toads (of any size). Unfortunately, a predator that tackles a large toad doesn't get a chance to learn; it's killed by its first encounter. This is why big predators die while smaller predators are resilient.

When invasive Cane Toads first arrive in an area, small marsupials, such as this Dunnart, try to eat the alien amphibians. Thankfully, however, these small predators soon learn to leave the toxic toads alone. Photo by Jonathan Webb.

Fifteen hundred kilometers (900 miles) farther south, Chris Jolly showed the same process at work within a single species. Yellow-spotted Goannas don't extend into these cooler regions; instead, the campgrounds are patrolled by Lace Monitors. A close relative of the famous Komodo Dragon, the Lace Monitor is the undisputed boss of every picnic area in New South Wales. Chris's surveys showed that as soon as Cane Toads arrived, the biggest goannas died. These 2-meter-long (6-foot) dragons have voracious appetites, so they grabbed toads and were poisoned by them. But Lace Monitors did not disappear entirely after toads invaded; instead, the big lizards were replaced by juveniles. Because young Lace Monitors are shy, and less inclined to bite first and think about it later, they are less vulnerable to Cane Toads. Before the toads arrive, young goannas hang out in the forest instead of the open campgrounds (to avoid their larger relatives). After toads wipe out the big bullies, the smaller goannas come in from the bushland and proudly swagger around the picnic tables previously dominated by their parents.

To look in more detail at goanna behavior, Chris got a few plastic trays from a fast-food restaurant. On each tray, he laid out a dead frog and a dead toad (with poison glands removed), plus a chicken neck as a control (to verify that the lizards were hungry). Goannas in toad-infested areas wouldn't look twice at a toad, but they would happily eat frogs. Goannas in toad-free areas ate both kinds of amphibians, but the toads made them sick, and they turned down a dead Cane Toad the next time it was offered. If you give them a chance to learn, these giant lizards can coexist with toads. Goannas are smart.

Not all native animals are capable of aversion learning. Tragically, our lab trials showed that my favorite animals—snakes—were bottom of the class. Some of the pythons and harmless (colubrid) snakes did okay, but it was curtains for the predator whenever a venomous snake (a species from Family Elapidae, like a King Brown or a Death Adder) met a large toad. Even after the snake ate a small toad that made it ill, it would enthusiastically seize the next toxic amphibian dropped into its cage. Evolution equips animals with the skills they need to survive, and it seems that Australian venomous snakes—perhaps because many of them eat only a few kinds of prey—have not evolved the ability to learn to avoid poisonous new types of food. As a result, taste aversion doesn't save the snakes. In a small species, the snakes just go on eating small toads and getting sick. For larger species, though, that poor classroom performance is a death sentence. Eventually the snake tackles a toad large enough to kill it.

Fortunately, aversion learning isn't the only way that a native predator can deal with the toad invasion. The other route is a longer one, with considerable death along the way, but ultimately it produces the same result. In "evolutionary rescue," the predator population adapts via natural selection, which winnows preexisting genetic variation within a population and builds up the frequency of "toad-smart" genes. In a genetically variable population, some individuals will be less at risk from toad invasion than others. The lucky survivors are those that are predisposed to ignore Cane Toads rather than trying to eat them, or are less vulnerable to the toad's poison, and so forth. Those resistant individuals are the only ones that survive after the toads invade. They breed with each other, and their offspring inherit the parents' toad-smart characteristics. As a result, the next generation is better able to cope with toads. Population numbers crash, but eventually the new toad-resistant genes enable the species to thrive once more.

The mechanism of evolutionary rescue is straightforward and uncontroversial: it is natural selection, Charles Darwin's great legacy. The problem lies in the time frame. Can adaptation happen quickly enough to matter? If not, the population will be annihilated. Surely, eighty years of exposure is too brief for snakes to have evolved to coexist with Cane Toads? But as I discussed when talking about rapid evolution in the toads themselves, biologists are now rethinking the time frame of evolutionary change. And Ben Phillips showed that it's not just toads that have evolved quickly. Their victims have evolved as well.

For his Ph.D., Ben worked mostly with Red-bellied Black Snakes, large and spectacular elapids that are common along the eastern coast of Australia—except in places where Cane Toads have invaded. In site after site, the numbers of Red-bellied Black Snakes have crashed as soon as the alien amphibians arrived. If evolutionary rescue was under way, we'd expect to see changes in the snakes that live in toad-invaded areas. And there's an easy way to look for those changes: compare the toad-exposed snake populations to those from areas where toads don't yet occur. Ben did that, and he found exactly the differences that he had predicted. About half of the toad-naive Black Snakes that he collected (from yet-to-be-invaded areas) readily ate toads, whereas none of the snakes from toad-infested areas would do so. This seemed to be an innate difference, not a result of learning. The snake's tolerance of toad poison also had increased: snakes from toad-infested areas were less vulnerable to a standard dose of toxin. And lastly, Ben found a difference in body shape: at the same body length, snakes from toad-infested areas had smaller heads. A snake's head size limits the maximum size of prey that it can eat, because snakes can't tear up their food into smaller pieces. Hence, a snake with a small head can't swallow a lethally large toad.

A few years later, Ben found the same two patterns in a follow-up study on Death Adders at Fogg Dam. This was the project where he caught Death Adders before the toads arrived, kept them in the lab for a few months, then implanted radio transmitters and released them after the Cane Toad invasion had spread across the floodplain. While the snakes were captive, he recorded their feeding responses to toads. About half the adders tried to eat toads, and those individuals were doomed after release: when we went out to follow the transmitter signal, we found the bodies of predator and prey lying side by side. But the other half of the Death Adder population was toad-smart: they refused to eat toads in captivity, and also in the wild, so they were safe. Just as Ben found in his Red-bellied Black Snakes, the surviving adders were the ones with smaller heads, and the ones that had refused to eat toads in our earlier trials in the lab. Even if a predator can't learn to avoid Cane Toads, then, natural selection can come to the rescue.

The discovery that predators were adapting to Cane Toads was an all-too-rare "good news" story about conservation. But when media outlets put out stories about our research results, the toad-busting community groups were horrified. They feared that the public would cease worrying about toad impact and, therefore, stop trying to eradicate the invaders. Too bad. In my view, scientists need to provide impartial information; hiding research results from the public is a recipe for disaster. So my media interviews were a tightrope act: yes, toads are a catastrophe—but no, it's not the end of the world. It's difficult to put across both sides of the argument when the interviewer pushes you toward one and not the other, or when you have less than a minute to explain the latest discovery. Even when I explained both the horrific impacts of toads and the likelihood that those impacts were diminishing through time, the final version of the interview that went to air often focused on only one of those messages. I decided to be relaxed about it. At least we were getting *some* scientific information about Cane Toads out to the public.

My confidence in the reality of evolutionary rescue was boosted in 2009, when I traveled to Townsville, in far north Queensland, and spent a few days with my new Ph.D. student, John Llewelyn, to look at potential study areas. Two images stand out in my memory, both testifying to the resilience of native predators. By then, I knew that the most toad-vulnerable species around Fogg Dam were a lizard (the Yellow-spotted Goanna) and a marsupial (the Northern Quoll). In the four years since toads had arrived at Fogg Dam, I had seen no quolls and only half-a-dozen live goannas, in places where I once would have seen many. But in the bushland around Townsville, Yellow-spotted Goannas were common, even though the alien amphibians were abundant also. Coexistence between toads and goannas was impossible at Fogg Dam, but both species flourished around Townsville. So, goannas *can* coexist with cane toads.

The second image came at night, as we drove along a forest road. The headlights illuminated a Northern Quoll on the road eating an insect, and beside it a large Cane Toad looking on serenely. It couldn't have been set up better if I had hired a TV crew to create a visual image of evolutionary rescue.

Beginning in 2010, John took on the challenge of finding out how the goannas of Townsville survive in a toad-infested world. Have they evolved physiological tolerance to the toad's poison, or have they evolved (or learned) not to eat toads? It wasn't an easy project. Goannas are difficult animals to catch, and hard to handle. If you put a wild-caught goanna in a cage, the lizard will be more interested in ripping your arm off than in cooperating with the research protocols. On the plus side, though, goannas become very tame in campgrounds. Success in stealing steaks and barbecued chickens encourages these giant lizards to treat humans with contempt. We took advantage of that boldness by focusing on campground goannas. John dangled either a dead (road-killed) frog or a dead Cane Toad in front of Yellow-spotted Goannas and recorded the lizards' responses. The results were clear-cut: Townsville goannas know all about toads and aren't interested in eating them. A frog, on the other hand, is always welcome.

The fact that Townsville goannas eat frogs but not Cane Toads fits the idea of evolutionary rescue. To be sure, though, we need to reject alternative explanations. In this case, it's possible that the goannas of Townsville had a preexisting aversion to Cane Toads (for some unknown reason), rather than evolving it after toads arrived. John found a way to test that explanation. The same goanna species occurs on nearby Lizard Island, which is toad free. Do island goannas treat Cane Toads as poisonous? No, they don't. When John dangled his dead frogs and toads in front of them, the island lizards tried to eat both types of prey. That result implicates evolutionary rescue as the reason why Yellow-spotted Goannas can coexist with Cane Toads in Townsville but not (yet) at Fogg Dam.

If we put the results from all these studies together, they tell us that even the vulnerable predator species can recover. The ecological impact of toads is most intense immediately following invasion. Given enough time, the native species adjust to the invader's presence and can live side by side with them.

You may have noticed something missing in my discussion of the ecological impacts of Cane Toads. So far, I've talked about pathways and mechanisms of impact, but I've said very little about changes in the abundance of wildlife.

Just how common are native animals after, compared to before, the arrival of Cane Toads in an area? Surely that is the gold standard by which we can measure the ecological impact of Cane Toads? Yes, it is—and Team Bufo spent a lot of time gathering those data. Because we needed long-term datasets, though, the surveys weren't completed until after we had completed our research on mechanisms of impact. The results of those surveys were humbling. Even when I understood the system in detail, I had failed to predict the overall outcome of toad invasion.

The only way to answer a complicated question is to divide it into smaller components. To examine the impacts of Cane Toads on a diverse assemblage of species, then, I began by looking at each species in turn, because we can't fit an entire ecosystem into a laboratory cage. The workshop at Middle Point was soon full of cages containing individuals of every native species I thought might be affected, to understand how meeting a toad for the first time affects everything from bugs through to birds. That information tells us about the results of *direct* interactions between toads and native wildlife. That is, what happens when a native animal comes face-to-face with a Cane Toad for the first time?

But that kind of study leaves out an important complication: native species interact with each other as well. That interconnectedness can massively change the effect of an invader. For example, let's consider an insect species that is eaten by Cane Toads. Will that species become less common after toads arrive, as the insects are turned into toad poo? Perhaps not. Cane Toads also kill off large predators (like goannas), so the insect may actually be safer than it was before. Cane Toads may gobble up fewer insects than were consumed by goannas in previous years. Hey presto—the numbers of insects increase rather than decrease. The same might be true for native frogs. From a frog's point of view, a Cane Toad is a booby trap that demolishes the frog's enemies. That benefit to Frogdom may outweigh any negative impacts of Cane Toads on native amphibians.

Because snakes were the focus of our work for many years before Cane Toads arrived at Fogg Dam, we now clearly understand their responses to the alien invasion. That has been a humbling experience. Ben, Greg, and I had published a paper in 2003 surmising that Australian snakes were in big trouble. Many of our native snakes eat frogs (and therefore would eat toads) and are very sensitive to the toads' poison. Thus, we confidently predicted that snakes would be killed in vast numbers and thereby become less common.

What actually happened? Eight years after we published that paper, and six years after Cane Toads arrived at Fogg Dam, we (the same three people) sat down to look at Greg's data on the numbers of snakes before versus after toad arrival. And it was demoralizing in one sense, because we had got it wrong; but uplifting in another, because the snakes had done better than we had feared. In fact, most snakes at Fogg Dam *increased* in abundance because of Cane Toad arrival.

How could that happen? How about floodplain Death Adders, for example? Surely their numbers will crash, especially given the specialized way that Death Adders obtain their food? They lie in ambush and wriggle the tips of their tails to mimic a small insect (the tail's tip is colored and shaped like an insect). Using a purpose-built machine to wiggle a tail tip from a road-killed snake, we found that the adder's lure is a magnet for hungry Cane Toads. In our lab trials, most toads that saw the lure hopped straight over and tried to eat it. Native frogs were more difficult to fool. Over the millions of years that Australian frogs have been exposed to Death Adders, natural selection has favored frogs that avoid the snake's lure. Today's amphibians are the descendants of frogs who ignored that enticing wiggle. Cane Toads have not had that evolutionary education, because there are no Death Adders in South America. I was confident that Death Adders would soon be rare on the Fogg Dam floodplain: suicide by tail-wriggling.

Ben's radio tracking in 2006 supported that prediction. Half of his radio-tracked snakes were killed within a few months of their release; we found them lying dead beside dead Cane Toads. Our nightly surveys along the nearby Arnhem Highway seemed to provide conclusive proof—adder numbers declined steeply over a three-year period after toads arrived, from an average of one snake per night to one snake every ten nights. Thus, we published a paper concluding that the toad invasion had been catastrophic for Death Adders.

But we realized that we were wrong as soon as we analyzed the data from Greg's nocturnal surveys. Although Death Adders were disappearing from the Arnhem Highway, they were more and more common on the wall of Fogg Dam. In the first ten years I had walked along that wall, I had never seen a Death Adder there. But now Greg and I encountered them frequently. Adders weren't the only ones to show that paradoxical pattern. Other snake species had also become rare along the highway but not elsewhere. Why? The answer was clear: snakes living near the main road were being killed by vehicles running them over, rather than by toads poisoning them. Even snakes that don't eat amphibians had become rare on the highway. So, there goes our "evidence" based on road counts.

But that still leaves the 2006 radio tracking, when many of our snakes died as a result of toad poisoning. How can Death Adders become more common at the same time that they are being killed by toads? The answer goes back to those indirect effects. Before Cane Toads arrived, Death Adders were being eaten by goannas (and King Brown Snakes)—the big predators that disappeared when toads invaded. So, although poisoning by toads reduced adder numbers, that effect was reversed by the removal of goannas. Around Fogg Dam, the numbers of Death Adders depend more on trucks and goannas than on the much-maligned Cane Toads.

Sean Doody's studies on the Daly River provide another example of goanna effects. Sean and his team measured rates of nesting and hatching by the giant Pignose Turtle in the Daly River. Until Cane Toads arrived in 2005, most turtle nests were robbed by goannas. The toad invasion was terrific news for turtles. Far more nests survived (about 20 percent more), because the toads eliminated most of the nest-raiding goannas. Freshwater Crocodiles obtained the same benefit, for the same reason.

Years later, Chris Jolly's surveys in coastal New South Wales in 2013 and 2014 provided a spectacular example of the benefits that can accrue to native species when Cane Toads wipe out goannas. If you want to know whether or not Cane Toads have arrived at a campground, look for Scrub Turkeys. These large, clumsy birds become abundant as soon as toads invade. Turkeys and their eggs (accessible in mound-nests at ground level) are easy prey for a goanna, so it's plain to see why the turkeys increase in numbers as soon as those helpful toads eradicate all the cold-blooded killers.

If I had to make a one-line summary about the ecological impact of Cane Toads in Australia, I'd say this: The goannas are massacred, and that crash in apex predators affects many other species. Which raises an embarrassing question—why didn't I predict this "trophic cascade" in advance? The sad reality is that I didn't understand ecosystem function well enough. It's difficult to comprehend the processes at work in a complex system as it operates in normal times. It's like peering at a car engine that is purring along smoothly. So much going on, so many interconnections. But if something goes wrong, and one component of the system stops working, it's far easier to see how that component affects other parts of the system. Cane Toads were the engine malfunction that, by taking out the goannas, showed us what role those giant lizards had been playing.

Before toads, my insights into the goanna's ecological role were obtained by dissecting stomach contents of a preserved specimen, or flushing the stomach of a live goanna with water so that it regurgitated (usually all over me). Our research at Fogg Dam provided many additional records of goanna feeding habits, either when we encountered goannas having their lunch or when our radio-tracked snakes became part of that lunch. Putting all of these observations together, it's clear that a goanna will eat almost anything. Inside the stomachs of preserved goannas in the Northern Territory Museum, I found everything from road-killed echidnas to goat horns.

But knowing that goannas have broad diets doesn't tell us what happens to the ecosystem if you take goannas out of the picture. It's the same problem the sugar scientists faced in the 1930s. Knowing that Cane Toads eat beetle pests in sugarcane plantations doesn't tell you whether or not toads control the abundance of those beetles. Common sense says that predation will cause a prey species to become less abundant, but that may not be true. For example, Water Pythons are abundant at Fogg Dam, and they feed mostly on native rats, so we might expect pythons to reduce rat numbers. But they don't, because the pythons mostly eat adult male rats. The males move around to search for females, a behavior that makes them vulnerable to an ambushing python. Meanwhile, the female rats are safe inside their soil cracks. More than enough males survive to fertilize all the females, so the numbers of baby rats in the next generation are not limited by the number of potential fathers. Pythons harvest the "expendable" males every year, with no effect on the number of rats that live at Fogg Dam.

So, even heavy predation sometimes may have no effect on the long-term abundance of prey. In other circumstances, though, even modest predation can have a major impact (for example, if predators target pregnant females). The only way we can find out if predators control prey abundance is to increase or decrease predator numbers, and see what happens to the prey. That's usually impossible (both logistically and ethically), but the Great Goanna Massacre, when toads came storming through the Fogg Dam floodplain, provided exactly that "experiment." Judging by what happened next (increases in abundance of smaller reptiles), those goannas had been eating enough of the local wildlife to control their numbers. We now understand that goannas are a keystone species—but we didn't know that until the toads pointed it out to us.

Once you recognize that Cane Toads can have positive as well as negative impacts on native wildlife, and indirect as well as direct impacts, many other results begin to make sense. For example, in 2012, Damian Lettoof looked at parasitic lungworms in Cane Toads and frogs along the New South Wales coast. We thought that toads might increase the numbers of frog parasites (by providing additional hosts), but Damian found the reverse. In areas with Cane Toads, the native frogs had *fewer* parasites. The reason for that perplexing result is that frog parasites are taken up by toads, but it's a one-way trip. The frog parasite can't survive inside a toad (because of physiological differences between toads and frogs). So Cane Toads (accidentally) hoover up frog parasite larvae that are waiting to infect frogs. End result: native frogs have fewer lungworms after Cane Toads arrive in an area.

One important lesson that Cane Toads have taught us is the difference between the fate of a few individual animals and the fate of the whole population. If we find dead animals lying in the bush after toads arrive, we're likely to interpret those corpses as evidence of an ecological catastrophe. That may not be the case. Over the years, we've found toad-killed predators of many species, but most of those species are still common. As in the case of the rats being eaten by Water Pythons, the important ecological issue is not the deaths of animals per se; instead, it's the impact of that mortality on the wider population. There is an important message here. Saving biodiversity is not about the fate of a few individuals. Remember that message the next time you see a marketing campaign by the local zoo, claiming that the birth of a cute monkey in Sydney will save the species in its disappearing forest habitat in Brazil. Conservation isn't about individual animals. It's about maintaining functioning ecological systems.

But let's get back to the theme of this chapter: the difficulty of predicting an ecological impact. Let's look at another set of complicating factors (in other words, another reason why I failed to predict the effects of toads). I assumed that if a snake eats frogs, it would also try to eat Cane Toads. That turned out to be wrong. I've already mentioned the goannas of Townsville, which readily ate frogs but turned up their noses at toads. Many other reptiles do the same. The ability to distinguish between (edible) frogs and (poisonous) toads is a critical skill for predators in many parts of the world. For example, many frog-eating Asian snakes know that the local toads are not edible. Those discriminating tastes are still present in snake species that evolved in Asia and have invaded Australia only recently (on an evolutionary time frame). A Brown Tree Snake or a Golden Tree Snake knows that Cane Toads taste revolting—hence, both snakes ignore them entirely or only consume small juvenile toads (that contain very little poison). These picky eaters aren't vulnerable to the Cane Toad invasion.

And to show how complicated it can all be, let's look at a bizarre case of geographic variation in the ecological impact of toads. This one takes first prize in the list of complicated, convoluted interactions that can arise in biological invasions. When Cane Toads first arrived at Fogg Dam in 2005, the Bluetongue Lizards were in trouble. These giant skinks eat any slow-moving prey item they can find, and toads are slow. The blueys around Middle Point, highly sensitive to the toad's poison, were almost exterminated. A lizard that ate even a small toad didn't live long enough to regret it. The same thing happened in Western Australia, as the toad front hurtled westward. But how about blueys in eastern Australia, where toads had been present since 1935? Had they managed to adapt to the presence of Cane Toads? Several decades of "death by toad" might have caused eastern Australian populations of the Bluetongue Lizards to evolve resistance to toad poisons. So, in 2008, we obtained some blueys from Brisbane and offered them small Cane Toads. The Brisbane lizards ate the toads and became ill—but, unlike individuals of the same species from the Northern Territory and Western Australia, none of the Queensland blueys died. They were many times more resistant to the toad's poisons. Were we seeing the rapid evolution of resistance, just as Ben had shown with his Red-bellied Black Snakes?

To check that interpretation, we obtained some blueys from the Sydney region for comparison. Cane Toads don't occur around Sydney (apart from one tiny population in an industrial suburb), so we didn't expect blueys from Sydney to tolerate toad poisons. But when we tested them, the ones from Sydney were just as resistant as the Brisbane lizards. How could this be, if the Sydney animals had never been exposed to toads? We have a likely answer to that question—and it's weird.

The explanation occurred to us one day in March 2009, when Ben, Matt, and I had taken a morning off from our discussions at Middle Point to go Barramundi fishing. Or, more accurately, we had moved our discussions of toad impact from the office to a boat, with a few fishing lures trailing behind us. And that conversation produced a possible explanation for the puzzle of Sydney blueys being resistant to toad toxin. Strangely, the idea involves a poisonous plant. About the same time that Cane Toads were brought to Australia in the 1930s, gardening enthusiasts brought in an ornamental plant from Madagascar. It's called "Mother-of-Millions" because of its rapid reproduction. Living up to its name, the plant spread through cities and towns across eastern Australia. It produces pendulous red flowers in winter and spring, and Bluetongue Lizards enthusiastically devour those flowers. Many lizards eat plants as well as animals—and blueys are highly omnivorous.

So, how does this relate to Cane Toads? The Mother-of-Millions plant produces toxic chemicals to discourage animals from eating it. It's poisonous enough to kill cattle. And, by an extraordinary accident of evolution, the plant's poisons are very similar to those of the Cane Toad. That's not as unlikely as it sounds; if a specific chemical offers an effective defense against enemies, it is likely to evolve independently within different types of organisms. That's what happened here. Because the plant is popular in the suburban habitats that are the stronghold of Bluetongue Lizards, the blueys of Sydney met Mother-of-Millions as they wandered around the gardens. They ate its flowers and they died. The only survivors were naturally resistant individuals, whose progeny inherited their resistance. That adaptive sweep might explain why the invasion of Cane Toads doesn't threaten blueys in eastern Australia. The lizard in your backyard is the descendant of great-grandparents from whom it inherited a genetically coded tolerance to toad toxins.

I'm not a hundred percent convinced that this is the right answer, but the evidence is strong. Blueys readily eat Mother-of-Millions flowers, and a bluey that can tolerate Cane Toad poison can also handle Mother-of-Millions poison. The only blueys that are able to eat a Cane Toad without dying are the ones that co-occur with the toxic plant. I wonder if the same kind of process might have created toad-toxin tolerance in other omnivorous species, like freshwater turtles and rats?

This peculiar example shows how a species' vulnerability to a new challenge depends on history. In the Mother-of-Millions case, an earlier invasion (by a garden plant) has changed the impact of a later invasion (by an amphibian). On a broader level, the colonization of Australia by European pastoralists changed outcomes for the Cane Toad by creating a landscape dotted with water sources, cattle dung, and artificial lights to attract insects at night. In ecology, everything is entangled. The impact and success of Cane Toads in Australia depend critically on earlier events, driven by people.

Of course, the toad invasion itself also has changed things. The most obvious changes are in the numbers of big predators (down) and their former prey (up), but it's more complicated than this. For example, the spread of Cane Toads through the Darwin area also provided a new host for some types of parasites. To see the full impact of the Cane Toad invasion, we need to step back. There are many pathways by which toads influence, and are influenced by, other parts of a complex tropical ecosystem.

When we take that broader view, there is no simple way to summarize the impact of Cane Toads. No native Australian species has gone extinct because of them, but a few species of large predators have taken an awful battering. Smaller species have been unaffected or have benefited as the toads removed the apex predators that used to eat them. There's no doubt that Cane Toads have changed the ecosystems they have invaded, and we'd be better off without them. But if I had to rate invasive species in terms of their ecological impact, I'd put humans at the top of the list. Next would come villainous aliens like Rabbits, Foxes, cats, and exotic grasses and diseases, which have all driven Australian natives to extinction. Cane Toads are bad news, but they don't deserve to be Public Enemy Number One.

So far, I've focused on the fate of native animals after Cane Toads arrive. But what happens to the Cane Toads themselves? As I mentioned earlier, the invasion front is composed of weird toads—large, frenetic, long-legged amphibians with low sex drive and compromised immune systems. They dominate the invasion front because they disperse faster than anyone else. But what happens after that wave of pioneer amphibians has passed through?

All across Australia, old-timers tell me that the first toads to arrive were "as big as a dinner plate" and that today's toads are tiddlers. Consistent with that story, toad-busting groups at the toad invasion front often trumpet their discovery of occasional outsized specimens. I am also told that toads "used to swarm in dozens under every streetlight," far more common than they are now. What can science tell us about such changes?

Unfortunately, not much. When toads first invaded places like Cairns or Brisbane, nobody bothered to measure or count them. Can we take those accounts by the old-timers at face value? Perhaps not. Ross Alford, an eminent amphibian ecologist at James Cook University in Townsville, has an interesting observation about the alleged hordes of giant toads in the good old days. A cruel consequence of studying Cane Toads is that everyone has a toad story to tell you—and, sadly, it's often the same story that someone else told you last week (or, if the person involved has had a few drinks, the same story *they* told you last week). Everyone assures Ross that toads have shrunk over the years. Everyone recalls the massive amphibians they saw when they first moved to Townsville, followed by a dramatic shrinking. The trouble is, some of those people arrived in Townsville thirty years ago, some only ten years ago, and some have only been there for a year. Have Cane Toads really continued to decrease in size across every timescale we consider? It certainly hasn't happened over the period when Ross has been keeping count.

Memory is unreliable. You are shocked at your first encounter with Cane Toads. You've never seen such huge amphibians. Yuck! But as you get used to them, the toads begin to look smaller. My first encounter with a glut of Cane Toads was around garden lights at a Fijian resort on my honeymoon. I was astonished at the toads' sizes—whereas these days, I wouldn't even look twice if I encountered the same scene.

For a long-term perspective, Ben Phillips measured thousands of preserved Cane Toads in museum collections. Average adult body size has decreased through time, but the shift is subtle. The invasion front is indeed dominated by big bruisers, but they were never the size of dinner plates. Personally, the biggest toads I've seen in the wild are in tropical Queensland, far from the invasion front. My student Dan Natusch found a 900-gram monster on Cape York. Over most of Australia, a "large" toad averages 300 grams, and a 500-gram toad attracts whistles of admiration from researchers—and snorts of disgust from everyone else. Based on Greg's counts, the maximum body size of toads at Fogg Dam hasn't changed over the decade since the toads first arrived, although average body sizes have fallen because of the production of juvenile toads (as the adults bred).

There's an occasional giant (or "snodger," one of Ben's favorite words) in every population, so reports of massive toads are most likely to come from situations where large numbers of toads are being captured. That bias explains why media reports of giant toads captured during "toad musters" mostly come from the invasion front. Enthusiasm for hand collecting toads is high, until people realize its futility. Vast numbers of toads are captured, and the biggest ones are paraded for the TV cameras. In contrast, people who have been living with toads for decades don't bother to collect them any more—or if they do, and they find that occasional "dinner-plate" animal, they unceremoniously pop it into the freezer. Big toads aren't newsworthy in Queensland. It's not surprising that those giant amphibians on the TV screen come from the invasion front rather than anywhere else.

In summary, it's a waste of time trying to explain why toads shrink in size behind the invasion front, because they don't (or, at least, not much). That leaves the decrease in toad abundance through time as the main pattern to understand. And oh boy, would we like to understand it! If Cane Toads become less common in the years after they first arrive, we need to find out why—so we could make it happen sooner. The implications for toad control are obvious. What's going on?

The data are weak, but they suggest that Cane Toad numbers do indeed peak a few years after invasion, and decline thereafter. As with so many things that involve Cane Toads, the best information comes from Greg Brown's nightly counts at Fogg Dam. The numbers of adult toads increased rapidly in the first two years after the amphibians arrived in 2005, but the plague numbers were maintained only briefly before toad numbers began to fall. Starting about three years post-invasion, toad abundance dropped by more than 90 percent over a period of less than three years, before recovering to about one-quarter of the peak level.

Importantly, Cane Toads aren't unusual in declining like this; most invasive species do the same. For example, Portuguese Millipedes were in plague proportions for a few decades after they invaded Adelaide, but these days they are scarce. So the process causing toad numbers to decline is likely to be a general one. There are three possibilities. The first is that the invaders eat themselves out of house and home. Like locusts moving through a wheat field, the hoppers at the front don't leave anything behind for the ones that follow them. That's possible but unlikely. There doesn't appear to be any shortage of bugs in tropical Australia.

A second possibility is that the fast-moving invasion front leaves its parasites and diseases behind, because infected toads are slowed down. The only toads at the front are the uninfected ones. Such a population might expand quickly, until the parasites catch up and the toad population declines. Most of the parasites carried by Cane Toads seem to have fairly subtle effects on their hosts, but that explanation remains possible.

The third possibility may be the most important one. Through time, native species work out how to exploit the invader. We're used to thinking of Cane Toads as invulnerable invasion machines—but, in fact, they have many Achilles' heels. For example, some species of native ants, bugs, snakes, rats, and birds eat toads. A buildup in these natural enemies may explain why toad numbers begin to fall a few years after the invaders first arrive.

If toads are so poisonous, how can anything eat them? Contrary to widespread urban myth, many predators can tolerate the toad's poison. For example, after an experiment in which he offered live Cane Toads to Mongooses in Hawaii in the 1940s, Charles Judd reported that five Mongooses had eaten twenty toads and were "alive, saucy, active and in excellent health."

We now understand how the Mongoose achieves this invulnerability. Bea Ujvari and Thomas Madsen have shown that even tiny changes in the sequence of a predator's DNA render an animal 3,000 times more resistant to toad poison. In genetic jargon, for example, Asian goannas differ from Australian goannas by only four base pairs in the DNA sequences that build the sodium-potassium-ATPase enzyme—but that small change makes the difference between "die in agony" and "look around for another toad for dessert." These same toad-resistance genes also occur in Australian snakes (like Keelbacks) that have close relatives in Asia.

More importantly, rodents have the same ability. When people think of Australian mammals, marsupials like kangaroos and quolls spring to mind. But, in fact, Australia has as many species of native placental mammals (like bats and rats) as of marsupials. Native rats are common all over Australia, and our cities are full of introduced Black Rats and Brown Rats. There's not much that a rat won't eat, and a spicy toad ranks high on the list of preferred menu items.

People often tell me about finding eviscerated Cane Toad corpses lined up in rows near the water's edge. What helpful (and orderly) murderer is responsible? Wading birds take out a few toads, but the mass killer responsible for these neatly arranged toad morgues is the large and formidable White-tailed Water Rat (Rakali). I've seen these giant rats killing Cane Toads at Fogg Dam, but I haven't yet done any detailed studies because (to be frank) I'm frightened of them. Weighing up to a kilogram (two pounds), they are quick, smart, and more than willing to perforate anyone foolish enough to harass them. Given the choice, I'd rather take my chances with a Taipan than a White-tailed Water Rat.

We have, however, done some research on the smaller native rats. At least, one of my braver students, Elisa Cabrera-Guzmán, has done the work—while I cowered in the background, trying to look dignified and professorial. Rodents bring out my inherent cowardice. Elisa found that native rats kill and eat even large adult Cane Toads. Years ago, when Thomas Madsen and I studied Dusky Rats at Fogg Dam (I wrote down the data while Thomas did the hazardous rat-handling), we estimated that the floodplain contained more than five tons of rats per square kilometer. That's a lot of hungry rats. It explains why Greg's radio-tracked Cane Toads avoid the floodplain, preferring higher ground where they won't end up as a takeaway meal for a killer rodent.

The native rat's city cousins are enthusiastic about the taste of *Bufo* as well. When Matt Greenlees radio-tracked Cane Toads in the Sydney suburbs in 2013, two of his toads were consumed by sewer rats. One of Australia's least

This radio-tracked Cane Toad met a grisly end when it encountered a native rat. Greg went to locate it, and all he found was a corpse—the rat ate the tongue and left the rest. Photo by Greg Brown.

popular invaders (the Brown Rat) was taking out another unloved alien (the Cane Toad). I didn't think that anyone would suggest breeding up Brown Rats to reduce toad numbers, but I was recently contacted by a man who wanted to do exactly that. He had discovered that the feral rats in his house were killing toads, and he planned to breed these super-rats to sell on the Internet as toad-destroyers. I'm dubious. How many people would prefer a rat plague to a toad plague?

In case you're not feeling sorry for Cane Toads yet, they are also under attack by an army of heartless murderers. And when I say "heartless," I mean it. Invertebrates (like spiders and insects) don't have hearts—and so the toad's poison doesn't give them a heart attack, as it would a goanna or a quoll. The floodplains teem with invertebrates big enough to eat young toads. The easiest ones to watch in action are the giant "Meat Ants" that ruin your picnic when they swarm all over your chicken drumstick. These ants are formidable: big pincers, plus a poisonous bite. They crisscross the edges of water bodies looking to scavenge anything that washes up, or kill anything too slow to escape them.

A month or two after Cane Toads lay their eggs, the tadpoles metamorphose into thousands of baby toads that swarm at the water's edge. The newly emerged toadlets are less than a centimeter (half an inch) long and weigh only a fraction of a gram. These miniature creatures live on a tightrope: they dehydrate on dry ground, and they drown in the water. So, in the harsh dry season of the tropics, the tiny toadlets can survive only in the thin, muddy fringe of the pond. Crammed into a strip of damp soil around the water's edge, it's a living carpet of tiny toads. Easy pickings for an ant with an appetite. There is nowhere to hide on the flat, open, muddy banks of the billabong. Bloodshed, carnage, annihilation: it's painful to watch, even if you don't like Cane Toads. A Queensland farmer told me that after Cane Toads arrived on his property, the Meat Ants shifted their colonies to the water's edge to exploit the hordes of metamorphosing Cane Toads every summer.

Ants aren't the only ravenous invertebrates out there on the floodplain. Elisa, the student who studied rats, looked at aquatic assassins as well (they breed tough señoritas in Mexico). Given a choice, water beetles (Family Dytiscidae) and water bugs (Belostomatidae) prefer toad tadpoles to frog tadpoles—probably because the small, slow *Bufo* tadpoles are so easy to catch. And juvenile toads are in peril from spiders as well. Given all these voracious murderers with a taste for amphibians, it's surprising that any Cane Toads survive to adulthood.

As I said earlier, those hordes of small hungry mouths may explain why toad numbers fall a few years after the amphibians first arrive in an area. For anybody that can tolerate bufotoxin, the Cane Toad invasion represents free meals. That food subsidy enhances the survival of toad-predators, so their numbers increase. Before long, the toads are surrounded by enemies. The honeymoon is over.

The message is encouraging. Australian ecosystems are not passive victims of the Cane Toad invasion, cowering in the corner while the amphibian bullies swagger around and kick dirt in the faces of the native wildlife. First, many native species benefit when toads knock out the big predatory goannas. Second, many species exploit the toad for food. In 2006, Ross Alford calculated that if all the toad eggs that were laid in Australia survived to produce adult toads, we would be knee-deep in the invaders. Ross estimated that only about one in 3,000 Cane Toad eggs makes it through to adulthood. Many are killed by desiccation, overheating, diseases, and cannibalism—but the native animals of Australia deserve some credit also. In many of our trials—even in large outdoor ponds—ants and Water Beetles rapidly eliminated all of the toadlets or tadpoles on offer. Every year, millions of baby toads end up as sashimi for Meat Ants or sushi for Water Beetles. And it's a volunteer army; they don't even apply for government grants.

What does this mean for the Cane Toad's overall impact? The benefits to Aussie wildlife have to be lined up against the ecological devastation that Cane Toads have caused, and it's not easy to calculate the bottom line. Cane Toads haven't caused the extinction of any native Australian wildlife species, but they *have* killed a lot of native animals. For a few species of large predators, it's been a bloodbath. And we need to see the toad's effects within a broader context. Wildlife populations are in steep decline across Australia, including in places where toads don't occur. Biodiversity is in free fall for a thousand reasons. Eliminating Cane Toads from Australia would help, but it would do little to combat the ongoing demolition of our unique wildlife.

SEVEN

Citizens Take On the Toad

> Those guys [Team Bufo] come up with the wankiest bull-shit about toads.
>
> GRAEME SAWYER, former lord mayor of Darwin,
> front page of *NT News*, March 17, 2013

Just as I was learning that my initial fears about the effects of the Cane Toad invasion had been overblown, public opinion was moving in the opposite direction—toward a higher level of fear and loathing. By 2005, when Cane Toads reached Fogg Dam, the species had become an icon of evil in Australian society. No other amphibian, anywhere in the world, had such a high media profile—and that public image was intensely negative.

The deep antipathy toward Cane Toads in Australian society has many sources, some obvious, others lying deeper in the collective psyche. Attitudes to toads are very different in other parts of the world. Toads of one species or another are found in almost any place that experiences warm to hot conditions (even Canada and Sweden, which is stretching the definition of "warm"). People must have encountered toads from the beginnings of human evolution. In Europe, Africa, Asia, and the Americas, anybody wandering down to a pond for water, at any time in the past few thousand years, was likely to encounter a warty little amphibian. Those meetings gave rise to stories in many different cultures, with the toad as likely to play the role of hero as of villain.

There were no humans around for a few million years after the ancestor of *Bufo marinus* evolved, but the first people to move into Central America would have rapidly encountered Cane Toads. Early interactions between people and Cane Toads were probably similar to those involving other toads. Coexistence was uneventful. Cane Toads sat happily beside their flooded ponds, and nobody took much notice of them. Having evolved within this system, the toad posed no problems to other species, or to humans.

But the relationship between people and toads has played out very differently in Australia. Long before Cane Toads emerged out of Queensland in

1983 and began marching north and west through the Northern Territory, Australian citizens had been treated to abundant negative portrayals of Cane Toads and dire warnings of an unfolding ecological catastrophe. But one of the seminal moments occurred in 1988, when filmmaker Mark Lewis released a quirky documentary about the amphibians and their effects on native animals. That film became the best-selling documentary in Australian history, and it influenced an entire generation. Most Australians are very fond of our native fauna, and Lewis's film played upon that affection to portray the Cane Toad as an interloper that, in threatening our unique wildlife, somehow threatened Australian identity.

Understanding people's attitudes toward Cane Toads is a critical part of the whole soap opera that is Toads in Australia. Fortunately, we have information—not just impressions—about that topic, because government researcher Rachael Clarke and her colleagues traveled around tropical Australia, asking people a standard set of questions. The answers were illuminating. In parts of Queensland where toads have been present for decades, the alien amphibians are unwelcome but are not seen as catastrophic. People have learned to live with them. Although nobody enjoys stepping on a toad when they take out the garbage at night, the Queenslanders know that there are still plenty of native birds and animals around.

In fact, some people have a sneaking respect for toads—and, occasionally, even a hint of affection. When Prince Charles and Lady Diana were married in 1996, the Queensland government presented them with a book bound in Cane Toad leather. A bicycle race in north Queensland is called Le Tour de Cane Toad. On New Year's Eve 2014, a Queensland radio station ran a "kissing under the mistletoad" competition, asking listeners to photograph themselves kissing their partners as they held a Cane Toad above their heads.

Cane Toads have a unique place in Australiana. Hidden among the torrent of abuse are a few light-hearted glimmerings. Small girls dress up toads in frilly outfits, and lonely old men enjoy the company of the amphibians that gather around the back door at night to feed on bugs from the overhead light. Talking to a toad as you slip into alcoholic insensibility is more fun than doing it alone. In recent times, author Morris Gleitzman cast a Cane Toad with an injured hind leg ("Limpy") as the hero in books like *Toad Rage*. Limpy is a likable toad, respectful of his forebears (he keeps the flattened road-killed husks of his aunts' and uncles' bodies under his bed). He can't understand why everyone hates him.

Outback balladeer Slim Dusty praised the alien amphibians in his song "Cane Toad's Plain Code," a celebration of the toad's prosaic, unhurried lifestyle:

> For there in boots and moleskin pants,
> He stood with proud and haughty glance...
> While humans fight with creed and race,
> And try to conquer outer space,
> I just plod on from place to place,
> And watch this mad old world go its pace.

But in today's faster-paced world, sober reflection isn't enough. We need our toads to perform, not spout words of wisdom. Cane Toad races are a staple of hotels from Brisbane to Cairns. On my honeymoon in 1978, I won ten dollars by backing the fastest toad (my most successful lifetime bet). Cane Toad racing requires substantial inebriation, especially if the punters have to kiss their toads before the race. Numbered toads are dumped in the center of the table, and the onlookers shout encouragement as the warty athletes head for the table's edge. First over the line wins a prize for its owner, and the toads are then returned to their bucket. The toads have no clear idea of what's going on, but nor do their supporters.

For a while, "Cane Toad" was a common nickname for any Queenslander, almost supplanting the revered "banana-bender." But it's dropped out of fashion again. These days, the Queensland rugby league team is called the Maroons, not the Cane Toads, but the oversized amphibians still hold a special place in the collective subconscious of the sunshine state. When Queenslanders voted to select official icons of their state in 2009, the Cane Toad rose to the top of the list. In areas where toads have lived for decades, people are as likely to use them for entertainment as for garden fertilizer. Nonetheless, the relationship is ambivalent at best, and it's rare to find a Queenslander openly admitting to toad admiration. Cane Toads are the villainous invaders that Aussies love to hate. When a toad wanders onto the back lawn, most people greet it with a shudder of revulsion, a scream, or a cricket bat. Toads looking for a warm welcome will be disappointed.

Closer to the toad invasion front, the opinions are far more negative. Resenting the recent arrival of Cane Toads, people blame the amphibians for any change in the local ecosystem. Even fiercer passions hold sway in areas that are soon to be invaded. Here, people are convinced that toads will kill off all the wildlife, and then eat domestic pets and children for dessert. The level of terror in soon-to-be-invaded towns is far greater than is warranted by the actual impacts of Cane Toads; it's as if Hannibal and his elephants were about to appear over the horizon. One widespread story among indigenous groups was that Cane Toads were giant creatures that ran on their back legs, attacked people on sight, and could outrun a horse. No wonder people were worried.

As the toads' invasion of the Northern Territories progressed, the media coverage ranged from the merely hyperbolic to the hysterical. Newspapers ran lurid scare stories, with the gold medal for scare-mongering going to the *NT News,* Darwin's brash daily newspaper. It warned that toads would poison the drinking water, and that schoolkids would abandon their studies to get high smoking dried toad-skins. A legitimate (and admirable) concern for native-animal conservation among the citizenry was transformed into an irrational fear.

Public enmity toward Cane Toads has been amplified by other factors as well. That the toads are large, warty, rotund, and poisonous has surely not aided their image. Their appearance makes them a good candidate for demonization. Further, Cane Toads pose an undeniable threat to pet dogs. The risk of a dog being fatally poisoned by a toad is highest for ferocious predators like terriers. Dogs of other breeds ignore toads, or learn to leave them alone after a nasty experience that involves a trip to the vet and copious irrigation of the mouth with fresh water. But a small hopping object is irresistible to a terrier. A vet in Brisbane, Matthew Reeves, recorded ninety dogs (and only a single cat) being affected by toad-poisoning, four fatally. Of the dogs, almost all were small terriers. Within a few months of the Cane Toad invasion in 2001, a vet in Katherine told me that there were no Jack Russell terriers left in town. As the proud servant of two pampered West Highland white terriers, I hate the thought of Cane Toads in the backyard.

There are many objective reasons to dislike Cane Toads and to want them eradicated—including a genuine concern for native fauna. But those reasons do not fully account for the degree of passion (and resistance to rational argument) exhibited by many toad-haters in tropical Australia. Other things are going on in the subconscious minds of these folks.

Because toads—and amphibians generally—have been symbols of evil in Western culture for millennia, Australians are taught from an early age to see toads in a negative light. The Old Testament, for example, warns us about

THE INTERFET FROG

In any war, the first casualty is truth. One typical piece of misinformation involved newspaper reports in 2008 of the discovery of Cane Toads on the island of East Timor. Australian troops had been stationed on that small island in the Indonesian archipelago as part of an international (INTERFET) operation to facilitate fair elections. Local people named the toads "the INTERFET frog" and claimed that they had been brought in by Australian troops. The alien toads were poisoning villagers. Hundreds of chickens were dying. Australia's most high-profile "green" politician whipped himself up into a lather of righteous indignation, demanding that we pay compensation to East Timor. But when journalists asked me to comment, I pointed out that nobody had actually identified the interloper as a Cane Toad. Sure enough, it proved to be a case of mistaken identity. A new toad species had indeed arrived in East Timor, but it wasn't the Cane Toad. It was the Black-spined Toad.

A smaller animal than the Cane Toad, the Black-spined Toad has been spreading eastward from the Asian subcontinent for decades, island-hopping through Indonesia. As a result, Black-spined Toads are turning up more and more frequently in Australia, often inside the boots of backpackers returning from tourist destinations like Phuket in Thailand. One newspaper article in 2015 reported six Black-spined Toads found in travelers' boots in a single week, prompting the memorable headline "Phuket, I'm Going to Australia."

The identification of the offending amphibians as Black-spined Toads means that Australian soldiers weren't the culprits. And as for the stories of village chickens being poisoned, the same myth circulated in the 1930s as Cane Toads spread through Australia, and it still crops up today. But whenever anyone tests the idea (as the sugar scientists did in the 1930s, and as Christa Beckmann did at Middle Point), the chickens gobble up small toads as fast as you can hand them over, and ask for more. The toads' poison doesn't worry poultry. The chickens of East Timor are safe.

invasive amphibians. In Exodus 8:14, God sends a plague of frogs to devastate Egypt, using language that would fit nicely into the script of a horror film:

> Behold, I will smite your whole territory with frogs. And the Nile will swarm with frogs, which will come up and go into your house and into your bed-

room and on your bed, and into the houses of your servants and on your people, and into your ovens and into your kneading bowls.

Throughout Western literature, from Shakespeare to Hans Christian Andersen, toads are the familiars of witches, the omens of bad luck, and are often used as metaphors for ugliness.

Cane Toads also stimulate latent xenophobia. In Australia, as elsewhere, immigration has caused deep anxiety in portions of the population. While many people are not shy about stating their dislike of "foreigners," others don't want to be seen as racist. They may be ashamed of having such feelings about people different from themselves. For these latter folks, the Cane Toad is a politically correct target for their xenophobic inclinations. In such commonly heard phrases as "Toads don't belong here!" one can often sense the whiff of something deeper.

In a similar way, Cane Toads serve as a convenient lightning rod for our frustrations generally. It is easy to blame the toad for deterioration of the natural environment (there is an undeniable link between the two), but the alien amphibian also becomes a scapegoat for many people beset by wider issues of grief, loss, defeat, and disappointment. One hears of Cane Toads being blamed for just about every problem in tropical Australia, with the possible exceptions of global warming, male pattern baldness, and erectile dysfunction. I suspect that someone, somewhere, has blamed Cane Toads for those problems as well.

Perhaps the widespread antipathy toward Cane Toads that developed as the amphibians flooded out of Queensland was not a bad thing, all things considered. The unwelcome invaders were having a devastating effect on many of Australia's top predators. And I have to admit that I benefited personally from toad-hatred—it made it easier for me to obtain funding, something of no small consequence in an era of increasingly tight budgets.

On a personal level, though, I was moving away from feeling that toad hatred myself. My relationship with the Cane Toad has been a journey from unreasoning hatred, to respect, to admiration. Increasingly, I felt no enmity whatsoever for the individual toad. The trajectory of my feelings is the opposite of that experienced by Reg Mungomery, the man who brought the toad to our shores in 1935. Reg began in awe of the toad's value for beetle control and had a hushed reverence for the animals. To Reg, Cane Toads rated somewhere between sex and chocolate on the scale of Good Things in Life. But the toads let him down. They failed as beetle-eaters, and their antisocial habits alienated his neighbors. Increasingly, the decision to import toads was lampooned. It was humiliating. In his later statements about Cane Toads, Reg sounds like a jilted lover.

But my journey took me in the other direction. I began to see the toads as victims of human folly rather than perpetrators of evil. I wish that Cane Toads hadn't been brought to Australia, but it is absurd to blame them for the ecological carnage they have caused. They didn't ask to be brought here. That shift in my attitudes inevitably flowed through to Team Bufo's research agenda. My initial perspective was militaristic. We were studying an outsider—an alien being—an enemy that we needed to crush. It was a crusade. But in time I came to embrace a simpler view, stripped of emotion. We were studying a giant amphibian, of interest in its own right as well as for its devastating impacts on Australian ecosystems.

As it became apparent that Cane Toads would spread through the Northern Territory and then onward through Western Australia, many citizens—inflamed by the dire predictions they read in their newspapers and heard on TV, and influenced at least subconsciously by anti-toad prejudices—decided they could not take the toad invasion lying down. Throughout tropical Australia, they formed citizens' groups dedicated to halting the toad invasion and eradicating the toad where it had already infiltrated. Their basic strategy was to organize groups of volunteers to go to a specified area after nightfall, collect as many toads as they could, and kill them. These events came to be known as "toad-busts."

The toad-busting groups had great success recruiting volunteers to participate in their nocturnal outings. It was a great night out—a chance for families to spend time in the bush, for communities to find common purpose. Mark Dapin describes the phenomenon well in his quirky book *Strange Country:*

> There is something about toad-busting, with its semi-official status, quasi-military material, frontlines and euphemisms (like 'busting' and 'controlling') that suggests a craving among gentle people to wear a uniform and fight a war, in these days when the only respectable warrior is an eco-warrior. And I cannot shake the feeling that the toad is so easy to kill because it is unattractive and foreign. I have no doubt that the impulse to wipe out toads to save native animals springs not from a desire to destroy the ugly but to protect the beautiful, but it seems unsettlingly convenient that it is so much fun.

IS THAT REALLY A CANE TOAD, OR IS IT A NATIVE FROG?

Before you kill that big amphibian you found in your backyard, you have to decide whether or not it really is a Cane Toad. Your decision will determine that animal's future, so you'd better get it right. How can you tell if it's a toad?

Australians claim expertise in true-blue national pastimes like riding surfboards, drinking beer, eating Vegemite, defeating English sports teams, and identifying Cane Toads. Sadly, our self-belief exceeds our ability. Telling the difference between toads and native frogs isn't easy. In 1971, as an inexperienced student, I found a beautiful frog on Fraser Island in southern Queensland. I used up a full roll of film photographing it. But when I showed those pictures to my frog-savvy colleagues, they laughed. I'd spent an hour taking pictures of a young Cane Toad.

A large adult toad is distinctive, but small toads resemble the adults of many frog species—including some that are endangered. Are native frogs being slaughtered under "friendly fire"? You bet they are! The problem isn't too bad for professional biologists or for organized toad-busts where at least one person knows how to identify toads. But for lone vigilantes the error rate rises dramatically. Ruchira and Nilu Somaweera set up a display of live amphibians at shopping centers in Darwin and asked people which animals were frogs and which were toads. The answers were wrong about one-third of the time. That's bad enough if the animal is a toad and is released in error. But it's a tragedy if a native frog is killed by mistake.

In southern Australia, that suspected "Cane Toad" behind the outdoor toilet is far more likely to be a native frog—because there aren't many toads around. At one point, some frog enthusiasts in Sydney set up a telephone hotline for people to report Cane Toads that they found in their gardens. That's not as silly as it sounds. Although the main Cane Toad front is hundreds of kilometers north of Sydney, toads often stow away in trucks and hitch a lift to the Big Smoke. Someone who found a "toad" in their backyard could phone the hotline, and an expert would come over to identify the animal and (if it was a toad) dispose of it. But the team discontinued their hotline after the first ninety-seven calls, because only five of those calls involved Cane Toads. All the rest were native frogs—plus one Bluetongue Lizard!

It's a widespread problem. Ruchira and Nilu also asked national park agencies across Australia about native species that had been brought into their offices, mistaken for Cane Toads. Of the thirty-six species on that list, some of the alleged "toads" were animals such as adult Green Tree Frogs—which, frankly, don't look at all like Cane Toads. Even the toad-busting groups have run into problems. In December 2014, the Facebook page of the Kimberly Toad Busters included a section on how to distinguish between toad tadpoles and frog tadpoles—but, unfortunately, one of their photographs was wrongly labeled. So, volunteer-based collection and killing of "Cane Toads" can be bad news for native frog populations. On fieldwork one night near Sydney, I encountered a group of cyclists standing excitedly around a frog that was being squashed underfoot by one of them. He proudly exclaimed that they had found a Cane Toad and were about to kill it. It was an endangered species of native frog.

Telling the difference between Cane Toads and native frogs is hard enough when you have the animal in your hand, but much more difficult when you are zooming down the road at a hundred kilometers an hour. An iconic scene in the cult-classic *Cane Toad* documentary of 1988 involves a Volkswagen Kombi van swerving wildly to hit toads on the road. (Actually, those little bumps on the road were melons, not toads.) If you ask drivers (as my student Christa did), most of them proudly claim to aim for toads whenever they see them. When we got those survey results, I was shocked—partly because so many of the victims would be frogs, and partly because amphibians hit by cars often suffer a lingering death. But Christa's follow-up studies reassured me. She set rubber models of snakes, frogs, and toads out on roads near Darwin and watched what happened. The models were rarely hit, because most drivers don't see small creatures on the road at night. So, although we say we run over toads on purpose, most of us don't actually do it—thankfully.

Toad-busting is socially acceptable hunting. The pursuit of wild creatures was once a mainstream occupation for humans but is now deeply unfashionable among most city-dwellers. Despite that political incorrectness, though, there is a fierce primeval joy to finding and catching animals. Ask any fisherman. Or snake-catcher.

Toad-busters readily concede the joys of camaraderie, of camping out, of sitting around the campfire at night talking with friends and having a quiet drink. And the Cane Toad is the perfect quarry for urban humans. It is the right size (big enough to see, small enough to grab); it sits out in the open; it lives close to where people live; it's slow-moving and reluctant to run away when approached; and it's safe to handle (despite possessing fearsome toxins). And murdering toads is seen as a valuable public service. Even children and old folks can join in.

Killing toads lets people feel that they are taking direct action against environmental degradation. This aspect of toad-busting's motivation was picked up by a young Norwegian anthropologist, Jon Rasmus Nyquist, who spent six months with one of the community groups (the Kimberley Toad Busters) in 2012. In his thesis, Nyquist talks of the angst of people living in the once pristine wilderness of the Kimberley, seeing the environment degrade year by year and feeling powerless to prevent the creeping loss of what they hold dear. Sarah Brett, a toad-busting stalwart in the early years, clearly articulated this sense of loss at a conference in 2007:

> I have lived in the Kimberley for sixteen years now, and the numbers of creatures I see, as well as the diversity of species I see as a vet, is reducing year by year. I see virtually no small lizards anymore, far less Owls and Tawny Frogmouths, far fewer Green Tree Frogs, less Magpie Geese and few Antilopine Wallaroos. This is but the tip of the iceberg. The country is changing.... I truly believe that our wildlife is disappearing.

And remember, this is well before the Cane Toads arrived in Kununurra. Nyquist observes that because the root causes of environmental decay—more people, changed fire regimes, feral pests and weeds, agriculture, climate change—are complicated and difficult to fight, people latch onto something over which they can have at least the illusion of power. Biodiversity loss was happening before the toads arrived, but the toad provides a symbol of the problem, and it can make you feel better—even if only for a little while—to find a scapegoat for that rage and heartbreak. Killing a toad may not help, but it's better than doing nothing.

Many toad-busting citizen groups sprang up after the Cane Toad became Public Enemy Number One in Australia. Three groups, in particular, assumed leading roles in the toad-busting movement. They saw toad eradication as being their core business and shared an approach to toad-busting that I will have more to say about shortly.

The oldest of the toad-busting groups is Frogwatch NT, which started life in 1991 in Darwin as an apolitical operation headed by ex-rangers and dedicated (as its name indicates) to frog conservation in the Northern Territory. Frogwatch NT began to morph into a toad-busting organization when one

Toad-busting is socially acceptable hunting and can be a lot of fun—as shown by the enthusiasm of these young indigenous children in Kununurra. Photo by Yuri D'Amico of Save the Children Foundation.

of its members, a teacher-cum-website developer named Graeme Sawyer, saw an opportunity in the looming tide of Cane Toads. Graeme assumed a leadership role in the organization. As the toads approached Darwin in 2005, Graeme's slick communication skills propelled him to prominence in the local newspaper, a platform he used effectively to inflame public concern over the ecological damage about to be wrought by the invading toads.

Another important group, the Kimberley Toad Busters (KTB), formed in Kununurra, Western Australia, in 2005. The toad invasion was still 800 kilometers (500 miles) away, but the KTB's charismatic founder and leader, Lee Scott-Virtue, pulled together a group of like-minded Kimberley citizens to fight the interloping amphibians before they could reach Western Australia. Instead of waiting for the toads to come to them, they went out in search of the toads. The KTB traveled to the Northern Territory for their first toad-bust in September of that year. The KTB was a worthy successor to previous crusades that Lee had spearheaded, on issues like environmental mismanagement by government authorities. Lee proudly traces her ancestry back to

Mary Queen of Scots, and through the early settlers of the West Australian colony. In *Strange Country,* Mark Dapin describes her as having "an air of hygienic efficiency about her, like a ward sister in a teaching hospital."

The members of the KTB were impressively fervent eco-warriors. Their well-orchestrated toad-busts provided media opportunities to show off participation by prominent sports figures, celebrities, and politicians, as well as Vietnam War veterans, family groups, backpackers, indigenous youths, and just about every other group within society. The KTB's policy of social inclusion was their proudest achievement.

The level of enthusiasm for toad-killing within the KTB was nicely illustrated by Lee Scott-Virtue when she combined her wedding ceremony in 2006 with a toad-bust. Unwilling to waste a weekend on something as prosaic as a wedding, Lee and electrical contractor Dean Goodgame combined the ceremony with some judicious annihilation of amphibians. They drove from Kununurra to the Northern Territory as part of their regular weekend toad-killing schedule. Surrounded by their faithful crew of toad-busters, they had a quick ceremony to tie the nuptial knot, then changed clothes and went out to massacre toads. Mark Dapin quotes Lee's fond reminiscence of how they got 777 toads that night but were too tired afterward for the traditional wedding-night marital activities.

The third toad-busting group, the Stop The Toad Foundation (STTF), also arose in 2005, and also in Western Australia—but 3,000 kilometers (2,000 miles) farther south, in the capital city of Perth. STTF took a more measured approach than the KTB but were well aware of the virtues of publicity. They attracted their own celebrity patrons and spokespeople, including high-profile literary and sports figures. Their strategy was to conduct a few massive toad-busts each year, rather than a continuous series of small efforts. STTF wasn't shy of the media, but it was more low-key than the KTB. If you turned on the TV and saw people scurrying around picking up toads, it was less likely to be the Stop The Toad Foundation than either the KTB or Frogwatch NT.

The idea of volunteer armies fighting a certified enemy of Australian wildlife was a big hit with the media. In the early days of toad-busting, documentary-makers flocked to the tropics to film the toad-busting groups in action. Video footage of kids, grandmothers, indigenous family groups, and celebrities wandering around at night picking up Cane Toads was a surefire hit. The spokespeople of the groups spouted doomsday predictions about toad impact and put forth optimistic assessments about the ability of ordinary Aussies to solve the

DERANGED IDEAS ABOUT CONTROLLING TOADS

Some ideas about toad control are even sillier than the claim by the mayor of Darwin (as reported in the newspapers) to have shot 22,000 Cane Toads with a laser-sighted rifle. And yes, I'm having trouble believing that I wrote that last sentence. But bear with me. One idea was to build a fence to stop toads, rather like an amphibian version of the Great Wall of China. The idea never got any traction, perhaps because anyone who has flown over the deeply dissected landscape of the eastern Kimberley dissolves into helpless laughter at the thought of building a fence across it, let alone maintaining that fence in the face of cyclones and floods. And remember, it only takes two toads to get to the other side (on the back of a truck, or intentionally translocated by some drunken idiot) for it all to have been futile. After a single romantic interlude between two fence-crossers, there are 40,000 toads on the wrong side of the fence.

Even the idea of a toad-proof fence seems staid compared to some of the loony emails I get about ways to eradicate toads. One gentleman wanted to build a low-flying helicopter equipped with a giant vacuum cleaner to suck up toads from the swamps. One of the sweetest ideas was from two schoolgirls who planned to come to Australia after they finished high school, catch all the toads, take them home to Brazil, and let them go.

I've also heard about extraordinarily complex traps, genetically engineering wildlife to render them invulnerable to toad poisons, and dropping thousands of toad-eating snakes from low-flying helicopters along the toad invasion front. Or would we rather fit lethally powerful microwave arrays to decommissioned Russian MIG fighter jets? And I've been told, in many a pub conversation, that we just need to bring in some predators from the toads' native range—perhaps jaguars?—to restore the ecological balance. Even Reg Mungomery's ghost would veto that one.

problem. The leaders of these groups drew lines on the map and discussed in military terms how their battalions would cut off the invasion and stop the toad incursion in its tracks.

One would think that with a common enemy to fight, the toad-busting groups would cooperate, share resources, and coordinate efforts. In the early days, that was true. Frogwatch NT's Graeme Sawyer even joined the KTB's first three toad-busts, to show them how to do it. But relationships between the KTB and the other two groups soon became acrimonious.

Given the small-town roots of the KTB compared to the capital-city sophistication of Frogwatch NT and Stop The Toad, and adding in the passion of KTB's founder, sparks were bound to fly. The KTB's philosophy was there for all to see at one conference in Darwin in 2008, when CEO Sandy Boulter presented the KTB's official "mission statement." It included intentions to "seize control of all issues relating to Cane Toads" and "defeat rival groups." Sandy's legal training and combative attitude were key (and effective) elements of the KTB's campaign to pressure political figures into demonstrating their support.

The KTB's relationship with government wildlife authorities in Western Australia was tense, but its greatest antipathies were directed toward STTF and Frogwatch NT. Those groups responded in kind. In July 2006, STTF released an "open letter" entitled "Frequently asked questions about the relationship between the Stop The Toad Foundation and the Kimberley Toadbusters." In that letter, STTF repudiated a set of criticisms and proposed that STTF and KTB sit down for some constructive discussions. But the letter's blunt and accusatory language only moved the groups further apart. To paraphrase: "You yokels in Kununurra are catching toads in the wrong places, and we won't hand over any funds to you until you can organize a proper invoice." I doubt that message went down well with Lee and her mates. The KTB decided to circle the wagons, and shoot at anyone who approached.

Several issues divided the groups, but the biggest flashpoint was money. Why did cash matter? Hopping into the car for a pleasant evening slaughtering toads sounds like cheap entertainment. But toad-busting has cost a lot of money, even though most of the labor is from volunteers: the KTB, for example, has received at least $2 million in government funding. My jaw dropped when I first visited KTB headquarters outside Kununurra. Their parking area was filled with four-wheel-drive vehicles and even a bus, all proudly bearing the KTB logo. And a nearby shed was neatly organized with shelves containing hundreds of toad-busting packs, to be handed to volunteers. These were no little back-of-the-garage operations; they were large, well-resourced entities. Far better resourced, in fact, than any of the research groups.

By 2007, the Western Australian state government was giving both STTF and KTB $12,000 a month. But in October of that year, the KTB's Lee Scott-Virtue complained in a newspaper article that it was owed money by STTF, and that "tax payers will get better value for their money if it goes to the

Kimberley group." STTF denied those assertions, and rival Perth newspapers ran contradictory stories in support of each group. Claim and counterclaim turned into an ill-tempered circus—which, of course, was loved by the media. Inevitably, the two main Western Australian toad-busting groups formed alliances with different political parties: STTF with Labor and KTB with the conservatives. Those divergent political links didn't do much for anti-toad solidarity.

Onlookers to the world of Cane Toads took a certain morbid enjoyment in watching the leaders of the community groups engage in character assassination of each other. But at the same time, it was dispiriting to see so much energy directed toward politics instead of toad control, and so much misinformation on the nightly news. The battles sometimes seemed to be more about power than about Cane Toads.

Public conflict between STTF and the KTB eventually convinced the state government to step in. The wildlife managers were tired of funneling cash toward unruly organizations that spent so much time attacking each other (and the hand that fed them) rather than just collecting toads. Unusually, the Western Australia government appointed someone with real expertise to conduct a review. Tony Peacock is a senior biologist who began his career as a researcher before moving into administration. (Tony's doctoral project was on diarrhea in pigs, an ideal background for someone who now talks to government ministers about science policy.) Tony's report on toad-busting was hard-hitting. "In conducting this review," he wrote, "the author has been astonished at the animosity between the two Cane Toad 'busting' organisations." This conflict, he wrote, "causes very obvious inefficiencies. . . . It is up to individuals to swallow their pride and work together or step aside."

My scientific career hadn't prepared me for the role I was about to be thrust into. Suddenly, I was in public conflict with people who held very different views to my own. My "opponents" were passionate, confident, and astute in using the mass media. I had naively assumed that my role as a scientist was simple: I had to discover what toads were doing and figure out how to stop them. If my research identified productive directions, people from other organizations (community groups, politicians, and so forth) would embrace those new approaches. We all had the same goal in mind—to conserve the Australian fauna—so we would cooperate with each other to find solutions to this conservation crisis.

Addressing toad control in the public sphere turned out to be very different: it was a world of politics, more combative and sectarian than I could have imagined. I soon learned that Cane Toads in Australia are a lightning rod for public passions about conservation. Scientists' commentary about toads attracts vigorous debate—and, often, strident criticism—from people in the general community. Some of those people are articulate and well informed, and some are zealots spouting nonsense. Some are driven by lofty ideals, others by self-interest. I learned that in this environment, science plays a less prominent role than I had innocently assumed.

Because a hatred of Cane Toads is integral to the Australian psyche, emotion often trumps empirical evidence when it comes to toads and what to do about them. Further, politicians recognize that the passion attached to Cane Toads is gold. By uniting the community in outrage, an imminent toad invasion enables an astute spokesperson to garner media attention. Graeme Sawyer is a case in point. His toad-related media stardom eventually propelled him into politics proper and landed him the job of lord mayor of Darwin. As Nyquist notes in his thesis, Graeme rode to the mayoral position "on the back of the toad." (In an interesting reflection on Northern Territory politics, Graeme won the vote after the previous mayor was jailed for using council funds to buy a refrigerator, women's underwear, a punching bag, DVDs, jewelry, and a Darth Vader mask.)

The high media profile of Cane Toads pressured governments into offering resources (including cash) to deal with the problem. In truth, most of the spokespeople who emerged onto the National Toad Stage were motivated by ecological concerns rather than their own career prospects or a desire for popular adulation. For a small minority, however, the Cane Toad invasion was a heaven-sent opportunity to put the boot to their enemies or tap into government funding.

The onslaught of giant amphibians created a perfect storm politically as well as ecologically. Public concern about toad impacts created opportunities for would-be political leaders, raised the stakes for the citizen-based toad-busting groups, and prompted exaggerated rhetoric from both camps that further fueled the public's fears. So, as the wave of anti-toad hysteria swept across tropical Australia in advance of the trespassing Cane Toads, so too did a wave of nonsensical discourse, misleading information, and outright propaganda—all of it, in my view, qualifying as "Bufobabble."

Politicians and journalists were guilty of Bufobabbling, but the toad-busting community groups raised this form of speech to a high art. That sounds harsh, but it's true. I've heard many statements by these people on radio and TV, and I've been with them at workshops. So it can't be put down to the media misquoting them. Spokespersons from some community groups really *did* keep on saying things that didn't square with the evidence.

As you might expect from their slogan, "If everyone was a toad-buster, the toads would be busted," the KTB put out the most outrageous claims. The group's interactions with the media and with politicians were based on the notion that the best defense is all-out attack. Like a sensationalist newspaper, the KTB's anti-toad fervor led it toward breathtaking hype. Sometimes amusing and sometimes infuriating, many claims in KTB bulletins and press releases didn't seem to be the result of careful fact-checking. And that's a polite way to say it.

For example, when I checked the KTB's Facebook page in November 2013, the organization claimed to have 8,500 volunteers. But the entire population of Kununurra is around 6,000, most of whom are not KTB members. Photographs of toad-busting teams in the KTB newsletters feature fewer than a dozen people, and many of them appear in every photo, even months or years apart. So how can the KTB have 8,500 members? The answer lies in how the number was obtained. Kununurra is a mecca for tourists. Many backpackers and older travelers ("gray nomads") pass through, and they enjoy the chance to go out one night looking for toads. So, that 8,500 is a cumulative total of participants. Most of them only went out for a single night and are now back home serving drinks, raising children, or teaching anthropology in Oslo.

The KTB's Facebook page also tells us how many toads they have killed: "close to 3.6 million adult toads (equating to around 330,000 tonnes of toad biomass)." Let's take this one number at a time: toad tally first, then combined weight.

The KTB puts its running scores up on Facebook, like a sporting team listing its victories. The grand total has grown considerably since the first KTB toad-bust in September 2005. The numbers increased at a fairly constant rate for the first few years, up to 253,000 in February 2009; that's about 1,500 toads per week, a number I can believe. It fits with my own experience in terms of the numbers of toads you can find, and it lines up with the tallies by other toad-busting groups.

But then the KTB's kill rate skyrocketed. By November 2010, the KTB was claiming 1,300,000 toad scalps, indicating an average monthly tally of 50,000 toads. And that number just kept going up. To get to 3.6 million adult toads by late 2013, the toad-busters were catching 70,000 toads every month. That's an average of over 17,500 adult toads per week.

Can those astronomically high numbers be true? The KTB claims to have marshalled 240,540 hours of volunteer effort to catch those 3.6 million toads. At that rate (about fifteen toads per person per hour), how many people would be needed to catch 17,500 toads per week? If the teams are out both Saturday and Sunday nights on every weekend, and each person catches fifteen toads per hour, you'd require many hundreds of people out there in the swamp every night. That's far more than rival STTF achieved for their well-publicized, once-a-year, mega-toadbust. And the KTB would have to achieve it every Saturday and Sunday night. I'm skeptical.

Average values can be misleading (the average reader of this book has one ovary and one testicle), but further analysis confirms my suspicion that the KTB leaders are not the best accountants in the world. For 3.6 million toads to equate to 330,000 tonnes, as the KTB claims, each adult toad has to weigh 92 kilograms (203 pounds) apiece. Those toad-busting nights around Kununurra must be real adrenalin-charged adventures!

Unfortunately, light-hearted humor is not a feature of KTB press releases. Instead, the KTB uses military terminology. The members call themselves an "environmental army" that sets up "reconnaissance teams" to map the "front line." Leaders have military titles like Campaign Field Coordinator, Safety Field Operational Leader, Operational Field Strategy Leader, Field Equipment Advisory Leader, Euthanasia Control Leader, and so on.

The KTB developed parochial paranoia into an art form. To call their communication style "antagonistic" or "combative" doesn't even begin to capture the passion of their prose. Their frequent newsletters and media releases drip with lurid colors, slogans in capital letters, inflated claims, and warnings of impending doom in a strident tone. Those newsletters remind me of Cold War communist rhetoric—but directed, in this case, at the perfidy of the government or rival toad-busting groups rather than the Capitalist Running Dogs of the USA.

If you are actively engaged in fighting for a cause, you become more and more passionate about it. That's human nature. If you're a toad-buster, tales of impending catastrophe wrought by toads give your cause even greater meaning. So, not only were the toad-busters in the trenches willing to believe prophecies of an ecological Armageddon due to toads, but their leaders were motivated to promote and popularize such ideas. The KTB website still warns of doom, saying the Cane Toad "will literally destroy one of the last unique biodiversity wilderness frontiers in Australia."

Not all the toad-busting groups veered into the realm of all-out propaganda the way the KTB did. STTF, for example, was careful about the information it put out. STTF's website talks about catching tens of thousands of toads (for example, 48,000 in 2005; 69,000 in 2008), not millions as the KTB does. The media statements by Frogwatch NT were also fairly low-key—compared, at least, to the KTB's passionate cries of outrage.

Toad-busting relies on two ideas. One: Cane Toads devastate native wildlife. Two: By killing adult toads, citizens can control and perhaps even eradicate the interlopers. By 2010, the results of our research were telling us that both pillars stood on very unstable ground. Cane Toads were bad news for some native species, but overall their impact wasn't as severe or as irreversible as initially feared. And killing adult Cane Toads—or trapping them, or trying to exclude them with fences—was futile.

Still not disabused of my naive beliefs about politics, I hoped that our growing knowledge of Cane Toads could better inform the public debate and move things toward a plan for rational, evidence-based action. I hoped to convince the community-group leaders that their collecting activities had as much chance of stopping the toad invasion as I have of being selected to play in the Australian rugby team, that the toads were not the Four Horsemen of the Apocalypse, and that citizens' desire for action could be channeled into more useful forms. That storyline would make a delightfully uplifting movie, but real life was different. The toad-busting leaders didn't want to hear what I had to say, and when I persisted in saying it, I was met with hostility.

The toad-busters shouted me down. The scientists were defeatists, they said; we were giving up before we even started. They, the hardy toad-busters, would prove that Aussie spirit could triumph over these bothersome invading amphibians. "At least we're doing *something*," the Toad Warriors exclaimed.

University-based scientists were branded the enemy, like rival groups and government wildlife authorities. As soon as we announced the results of a research project, our conclusions were challenged. When media stories appeared about research that Ligia Pizzatto and I had done on toad cannibalism, the KTB responded (in their November 2011 newsletter) that we were wrong. Cane Toads were not cannibals! The KTB had dissected thousands of toads and had never found other toads in their stomachs! Interestingly, for the first and only time, the KTB had sent me a draft of this newsletter before they put it out. I responded by explaining that the cannibals are the juvenile toads, whereas the KTB dissections have always focused on adults. I even sent the KTB a picture of a dissected juvenile toad with several smaller toads inside its stomach. But the newsletter was published without any changes.

In hindsight, I should have expected a hostile response to our research findings. After all, I was attacking the foundation of the toad-busters' reason for existence. I was telling them that toads were not as big a problem as we had feared, and that catching and killing adult toads—the activity on which many based their identities—was futile.

After a few months of beating my head against that particular brick wall, I began to understand what was happening. Community groups need the support of local people. The only thing that the toad-busting groups could do, until our research finally produced some powerful new methods, was to go out and pick up toads. Any community-group leader who admitted that this wouldn't have any effect would be out of a job. And so they clung fiercely to the notion that toad-busting was effective.

Holding on to a belief despite contravening evidence is something we all do, but in this case it represented a real clash between scientists and toad-busters, a divergence in worldview. Every scientist I've ever met has scoffed at the suggestion that killing adult Cane Toads can have any overall effect on toad numbers, but this is exactly the activity that most of the community groups espouse. How can well-meaning people reach such diametrically opposed conclusions?

It's easy to see why somebody might think that removing toads will help. After all, it works for other kinds of animals. If people shoot every Koala that they see, Koalas will become rare. But Cane Toads aren't Koalas. Physical removal of adult Cane Toads doesn't work because it can't keep up with the toad's incredible reproductive output. *For toad-busting to reduce toad numbers, we have to remove toads quicker than they can replace themselves.* That's not a huge hurdle with a species that produces only a few offspring each year. It's why the Koala was almost exterminated by hunters. But with a species that produces tens of thousands of offspring per clutch, you're wasting your time. Any adults you remove will soon be replaced by the next generation. Mathematical models tell us that unless we can remove almost all of the adult toads on a regular basis, physical removal has no effect. None whatsoever. And sorry, guys—it isn't possible to capture that many toads.

The other issue earning scientists the ire of the toad-busters was the ecological impact of the Cane Toad invasion. If (as we were discovering) the toads are not such a massive problem, the toad-busters' activities are less important. As I gave media interviews about Team Bufo's results, saying that toad impacts weren't as disastrous as we had feared, the toad-busters took aim at me. Whenever a research story hit the news, the toad-busting honchos looked for a reason to pooh-pooh it.

The pattern was consistent. The first day's press reported what I said, with a positive spin. Then someone contacted Graeme Sawyer for a response, and he'd either say that yes, this was true, but he knew it long before I did; or no, the research was nonsense. He said the same things about everyone else's research too, so it was too predictable to be off-putting. I looked forward to Graeme's colorful reactions. And, in truth, it was pretty tame stuff compared to the wars between the competing community groups. Scientists were not major players in the media battles, probably because most of us don't have the larger-than-life persona that loves to be front and center in a knock-down, drag-out public battle.

Despite that weakness, we scientists held our own in some of those media merry-go-rounds. Unlike the toad-busters, we had hard evidence on the impacts of toads. Information is a powerful weapon. Because of Greg's long-term monitoring, I could say things like "There are twenty-three percent more snakes at Fogg Dam now than there were before the toads arrived." That's hard to argue against. If the toad-busters responded with apocalyptic nonsense, they began to sound like Chicken Littles running around screaming that the sky is falling in. But that was the reaction of only part of the public. The rusted-on supporters of the toad-busting groups decided that I was an idiot, or worse. Before long, I was accused of being a Toad Apologist with all kinds of murky motivations. As far as I can recall, though, nobody ever accused me of actually taking bribes from the toads.

One of the strangest aspects of the Cane Toad Wars was the way that evidence was discussed when scientists and toad-busters came together in person. And we did come together from time to time, because it was obvious to the government (which was paying the bills) that there was a major inconsistency between the messages being spread by the two sets of people. Surely these could be reconciled if we sat down together and talked it over? That impetus led to several workshops where toad-busters and scientists talked about our results and plans. When I first accepted an invitation to one of these events, I expected fireworks. My presentation was peppered with conclusions like "Toads aren't a problem for most species of native wildlife" and "Physical removal of adult toads has no long-term effects on toad numbers." I expected to be shouted down or dragged off the stage—or, at the very least, abused in the corridor afterward. But the toad-busters politely ignored my conclusions. Perhaps they felt that I had a home-ground advantage at a scientific workshop, and they were more comfortable in a media interview. Whatever the reason, there was no serious debate. The toad-busters' own talks described their activities and plans but made no claims about their impacts. They avoided the overblown statements that filled their public speeches. We might as well have been in parallel universes.

The public debate changed as the toads completed their takeover of the Northern Territory and began to spread through the Kimberley. You only had to go outside in the evening, and see toads gathered around the porch light, to recognize that all of the brave talk had come to nothing. The toads had poured through unhindered. Despite the optimistic media releases and exaggerated claims, it was clear that toad-busting was having only a trivial effect. Cane Toads were flooding westward like an amphibian tsunami. Even the most zealous KTB comrade must have seen that the writing was on the wall. Nevertheless, the community-group spokespeople continued to bluster and to criticize the scientists' conclusion about the futility of hand collection as a means of toad control.

When the federal government released its long-awaited Cane Toad control strategy ("Threat Abatement Plan," or TAP) in 2011, the news got worse for the community groups. The TAP concluded that toad-busting had minimal effect. Given that the toad invasion was expanding just as fast as ever, that conclusion was inevitable. But there was an even worse problem for the community groups. In future, the TAP recommended, the allocation of funds for toad control should be based on results, not press releases. To obtain public money, you need to show that your activities control toads. The toad-busters had flourished because they were effective politicians, but now the rules had changed. To maintain their income stream, they had to show that they were reducing toad numbers. It was an impossible task.

With the federal government now out of the picture, the community groups redoubled their pressure on state and territory governments for continued funding. This worked for a while, but political interest was waning, and the media releases by the toad-busting groups ended with heartfelt pleas designed to wring more money out of the public purse.

The biggest challenge to community-group rhetoric wasn't opposition from scientists or shifting political winds, or even the much-despised TAP. Instead, it was the simple fact that the toad invasion just kept on coming. As the years rolled on, so did the toads. In the early years, the public rhetoric of the community groups was full of confident statements like "We *will* stop the toads!" But such words increasingly seemed naive. Indeed, the very name "Stop The Toad Foundation" began to look embarrassing. Recognizing the awful truth, STTF quietly disbanded in 2015.

As the toads' victory over the toad-busters became obvious, the earlier public acclaim for community groups began to be punctuated by grumpy complaints. Were the toad-busters just wasting public money? In response, the KTB and Frogwatch NT changed their rhetoric. They no longer talked about stopping the toads. Now the headlines read "We can *slow down* the toad invasion." And when that seemed too sanguine an assessment, the message evolved into "We can reduce the abundance of toads in local areas." The public didn't seem to notice the change in aspirations. People were losing interest.

And it just got worse, as the residents of Darwin saw more and more flattened toads on the road as they drove to work. In spirited media appearances, the toad-busters argued that if not for their activities, toads would be even more common. Graeme zoomed on, regardless of the mounting piles of toad carcasses by the roadside; he's the epitome of a man who sees the glass as half full. Although the toads were common even where his team had been toad-busting for years, he still brimmed with confidence that it was all going well. He pointed out that toads were less common in some areas than in others, and he applauded these zones as evidence of toad-busting success. Skepticism among the general public was now reaching epidemic proportions, however, so the message evolved again. Now the toad-busters were trying to "keep toad numbers under control until scientists could work out a solution to the toad problem."

At first, I thought this change in rhetoric was a good sign. The community groups, having recognized the futility of manual collection, were turning to us (the scientists) for help. Finally, toad control would be guided by research! But my optimism was short lived. Rather than handing over power, the toad-busters set up their own research projects. Academics like Team Bufo couldn't be trusted, so the locals would take over the research as well.

Unfortunately for the community groups, good research is not easy to do. It takes a great deal of organization, training, and clear thinking. Assigning a particular cause to an observed effect with a high degree of confidence requires a rigorous process. You need to measure what's going on, then test alternative possible explanations for that trend. To eliminate the possibility that some other factor is causing the effect, your data need to be collected carefully and from multiple locations. And you need to apply statistical tests to be sure that you are actually seeing a signal in the data (as opposed to just noise).

The toad-busting groups developed great skill in collecting toads and in obtaining media exposure and resources for their activities. But they ran into trouble when they tried to form research teams. Setting up a research enterprise is hard enough for a university. It's much more difficult for a small community group. Doing research is a full-time job. So, when the community groups tried to do their own research, it was a shambles. They had very little idea of how to go about it, no training in experimental design or statistical analysis, and a poor understanding of the background information in earlier papers.

Predictably, the results were dismal. None of the toad-busters' "research" has been published in the mainstream scientific literature. Team Bufo's work was possible only because of backing from the Australian Research Council (ARC), a government body that hands out funds for research. Success rates for ARC grant proposals are very low—around 10 or 15 percent. No matter how enthusiastic they are, a group without any scientific background has no hope of getting funding from such a scheme. And of course, the leaders of the KTB and Frogwatch NT didn't approach professional scientists for advice, because it would have been an admission of ignorance. If anything, the community groups became increasingly antagonistic toward university-based research programs.

Sadly, we scientists made limited headway in getting our messages across to the public. Even when our arguments were watertight and the data were overwhelming, many members of the general public continued to believe propaganda instead. In turn, that situation reflects the gap between professional science and the rest of civil society. Scientists work in an orderly world where the rules are clear, even when people disagree. Scientific inquiry is about finding truth, not winning an argument. Our conclusions need to be supportable by facts. If the evidence is ambiguous, we need to say so. Nonscientists don't feel the same way. Confidence carries the day, even if that confidence is based on self-belief, not evidence. Misleading pseudo-facts can be more effective than complex explanations.

The pointy end of Toad Politics was an eye-opener for me, because I used to see science as playing a central role in conservation. My undergraduate lectures featured inspiring examples of how field biologists had identified threatening processes, thereby saving endangered populations. If biologists could discover the best approach, the managers and general public would fall in behind. Everyone would live happily ever after. Unfortunately, conservation doesn't work that way. I started questioning the science-centered model when I conducted research on giant pythons in Sumatra. Hundreds of thousands of these snakes are taken from the forests each year, and (remarkably) it appears to be a sustainable harvest. But when I talked with policy makers in Jakarta, science played second fiddle to social, cultural, and economic issues.

Toad Politics was worse, because it was happening in my own country. It was disillusioning, and I began to doubt the truth of what I saw on the television news. Could a suave charlatan offer himself (or herself) up as an "expert" on some topic, and preach their bogus "research" conclusions on national TV? Sadly, the answer is a resounding "Yes." The journalists I meet are faced with impossibly short deadlines. It's a twenty-four-hour news cycle, and yesterday's story is dead and gone. A journalist's main job is to put together an entertaining take on the latest tale. Few have the time or background to weed out the evidence-based conclusions from the utter bullshit.

It wouldn't worry me too much if these problems applied only to stories about Cane Toads. But the same issues apply to other topics as well, including some with much greater ramifications than an American amphibian romping through tropical Australia. And unlike the toad-busters, some pressure groups are driven by cash or bigotry rather than conservation concerns. Climate change is bringing with it a holocaust that makes the Cane Toad problem look like a fart in a bathtub. I used to watch the TV news avidly, but nowadays I'd rather curl up with a good crime novel. For a scientist, a world where evidence is trumped by populist proclamations is deeply concerning.

The broader involvement of citizens in collecting scientific data, controlling invasive species, and restoring degraded ecosystems is one of the brighter developments of the early part of this century. "Citizen science" is vital not only in slowing the decline of biodiversity, but also in ensuring the continued viability of ecological processes that provide humans with food, clean water, and other necessities. Scientists and government authorities can't do it all, and engaging with the nonhuman world around us brings many benefits. But we also need to be aware of the potential pitfalls of citizen science as we tackle the myriad problems threatening our local and regional ecosystems.

EIGHT

The Quest for a Way to Control the Toad

> The clever men at Oxford know all there is to be knowed. But they none of them know one half as much as the intelligent Mr. Toad!
>
> **KENNETH GRAHAME**, *The Wind in the Willows*

It's all very well for scientists to learn about Cane Toads and write scientific papers (or even a book) about them. But whenever I talk to nonscientists, the conversation turns to a single issue: How can we control the toad invasion?

Eliminating toads has proved to be more difficult than people expected. Cane Toads are a formidable enemy because of their mind-bogglingly large clutches of eggs. It doesn't take an Einstein to realize that if two toads can produce 40,000 eggs in one clutch, picking up adult toads can't have any long-term effects. And the high reproductive rates of Cane Toads aren't exactly a secret. In a 1995 episode of the animated TV show *The Simpsons*, Bart's escaped pet (clearly based on the Cane Toad) multiplies so rapidly that it overwhelms Australia.

Unfortunately, Bart's clear view of this issue hasn't filtered through to Australians. It's not only community groups that have struggled; some professional biologists have had trouble with it also. The way we organize wildlife management activities hasn't helped. Scientists who work with insect pests are usually part of an agriculture-related division, in a different building from scientists who work with terrestrial (land-based) vertebrate pests. To control feral pigs or foxes, vertebrate-pest specialists often slaughter adult animals. But if the noxious pest produces thousands of eggs at a time (like the Cane Toad, and many insects), that approach is a waste of time. A fixation on how we can catch and kill adult toads is a guarantee for failure.

Nonetheless, that approach has been the mainstream effort. Australians have been bashing Cane Toads over the head with golf clubs ever since 1935,

The vast numbers of eggs produced by female Cane Toads mean that we can never control toads by culling adult animals; they can reproduce more quickly than we can remove them. Photo by Greg Clarke.

but no amount of clubbing, bashing, and running-over on the road has dented the toad's increase. A simple mantra, "If everyone was a toad-buster, the toads would be busted," inspired massive toad-culling exercises as the Hawaiian immigrants hopped westward. We now know more about the effects of different toad-busting methods, and I'll talk about the main ones below.

The most popular approach to Cane Toad control is straightforward. Cover yourself with insect repellent, go out at night, look for toads, pick them up, and kill them. All over Australia, the initial toad-busts were focused on hand collecting. And it can be quite effective—adult Cane Toads like to sit out in the open on suitable nights. Of course, sometimes they are tucked away in shelters and toad-busting is futile.

To streamline their activities, toad-busting groups turned toad reconnaissance into a sophisticated pursuit. On a dry night, wandering around just uses up your flashlight's batteries to no avail. One ingenious solution is to use someone else to find the toads. Who? Man's Best Friend. Specially trained "sniffer dogs" are used in many wildlife-management projects. If I needed to find dead birds lying around a wind-farm rotor in tussock grassland, I'd ask a dog to find the carcasses for me. And I'm confident that with the right dog, it would work.

But a sniffer dog is no use for hunting Cane Toads. When the toads are abundant, the habitat is saturated with their scent, and finding a few extra toads won't help anyway. If toads are scarce, there may be value in finding those few widely scattered individuals; but unfortunately, dogs aren't much use. The few sniffer dogs I've gone out with were better at finding dog poo than Cane Toads. One of them did manage to find a native duck and kill it. Another found a (planted) toad but immediately grabbed the toad, was poisoned, and barely survived.

Toad-sniffing dogs play well in the media, though. The immaculately coiffured TV presenter gazes adoringly into the eyes of this canine eco-hero and explains how he or she will save the planet from rampaging amphibians. Dogs like Nifty became celebrities, trotted out for TV shows from Kununurra to the New South Wales coast. But publicity value aside, it's an unproductive exercise. If you calculated the cost of training and dog food, and set it off against the numbers of Cane Toads found, every sniffer dog has died in debt. That dog may have a folder full of press clippings featuring its photo, but it will have very few toads to its credit.

Most people soon question the value of going out, night after night, to pick up Cane Toads. The backyard is still full of them, and your beloved life partner is complaining about the bags of toads taking up space in the freezer, awaiting a cold death and an eventual burial. In tropical Australia, space in the freezer is a valuable commodity. More toads mean less space for the frozen mangoes. And traipsing out every night to kill small warty things, when the aforesaid life partner wants to talk about the social entanglements at his or her workplace, creates marital tensions.

In a desire to salvage their love life and spend more time with the kids, many people have opted to install a toad trap that keeps working by itself 24/7. These traps are variations on a simple theme. Bright lights attract bugs, bugs attract toads, and a hungry amphibian ends up passing through a one-way door, or falling into a pit through a counterweighted trapdoor. Under the right circumstances, these traps can catch a lot of toads. But usually they catch very few.

Toad traps have been in use for a long time. One old farmer told me about a very simple version: a 44-gallon drum sunk into the ground. Small lizards and frogs fell inside and died, their bodies attracted flies, the flies attracted Cane

Toads, and the toads couldn't get out once they fell in. In time, their bodies acted as additional fly-attractants, and so it went. It was effective but nonselective, and it inflicted a lingering, inhumane death on the animals concerned.

The motivation to design a better trap got a boost in December 2005, when the Northern Territory government offered a $15,000 prize for the most effective toad trap. They received 114 entries, including one from Germany. An expert panel chose the most promising designs and ran a field trial. The clear winner was a trap with a drop-away floor under a bug-attracting light, built by Paul Baker, a diesel mechanic and mango farmer from Katherine. In the standardized field trial, Paul's trap caught forty-nine more toads than the runner-up. The second-best trap used one-way mesh gates that could swing inward to let a toad in but prevent it from exiting. In a triumph of marketing skills over capture rates, it was this second design—from the fertile brain of Graeme Sawyer—that took over the commercial market. For a while, it seemed as though there was a toad trap on every lawn in Humpty Doo. The Northern Territory government offered a thirty-dollar cash rebate (out of the wildlife department's budget) for anyone who bought a trap. Wildlife officers were bitter because money for conservation was tight, and the traps didn't really reduce toad numbers.

We tested Graeme's traps in 2006 and 2007 and came to the same conclusion as the wildlife department staff. We knew (from hand collecting, marking, and releasing) how many toads were around, and the traps caught only a tiny proportion of the local toad population. Many people who bought traps had the same experience. On talkback radio in Darwin, people complained that they caught very few toads and sometimes caught native animals instead.

Nonetheless, marketing trumps evidence. The Western Australian wildlife management authority invested in some traps when the toads finally crossed the border on their westward invasion. Fortunately, though, the professional toad-killers of Western Australia kept detailed records of the traps' effectiveness. On most nights, the traps caught nothing. On a good night, a trap got around three or four toads. Hand collecting and cruising the road were far more effective. And the traps had a serious downside as well. They caught feral cats (which is good ecologically, but worrying in terms of animal welfare), Blue-winged Kookaburras, Bluetongue Lizards, dragon lizards, Water Pythons, and even Freshwater Crocodiles. A native animal squashed in with a bunch of unhappy toads didn't survive for long.

If you can build a better toad-trap, the world will beat a path to your door. So, somewhat belatedly, scientists took up the challenge. At James Cook

University, Lin Schwarzkopf ran experiments to check the effects of trap design on capture rates. The results were illuminating. For example, white lights (as used in all of the traps) repelled toads! Switching to a "black light" (UV) made the traps ten times as effective. Adding a loudspeaker that played toad calls (especially baritone rather than soprano calls) attracted female toads—an important advance, because it's the adult females that we need to target for population control. Lin's new and improved Toadinator is far more effective than the earlier versions. But once bitten, twice shy. It will take a heck of an advertising campaign to encourage the citizens of Darwin to part with any more cash for Cane Toad traps.

If traps don't work, how about fencing waterholes to cut off the toads' access to water? The KTB condemned fences as an abomination, but Frogwatch NT hailed them as the new wonder weapon, able to devastate toad populations overnight. When Graeme Sawyer was extolling the power of fences to eradicate toads, he sounded like a fundamentalist preacher. That divergence in opinions generated more heat than light. In the *NT News* of April 16, 2008, in an article headed "Toad Feuders Hopping Mad. NT and WA environmental warriors fighting for territory," Graeme is quoted as saying that the leaders of the KTB are "off their tree at the moment, grasping at straws. They are actually doing quite a bit of damage to the Cane Toad cause."

With delightful symbolism, fences divided the groups. A kilometer (half a mile) of STTF's fences mysteriously burned down in September 2007 (the report says "destroyed by illegally set wildfire"), and diatribes by KTB leaders in local papers attacked the fences' impacts on native wildlife and lampooned their effectiveness at catching toads. It was emotive stuff. In a media release, the KTB wrote that "it would be a very sad day for the Kimberley to see people in their backyards sitting safely behind a Cane Toad fence, while our wildlife is being obliterated around them."

But the evidence says that fences can play a useful role at a local level. In a rare example of collaboration between scientists and community groups, a student at Sydney Uni, Dan Florance, investigated the effects of fencing off artificial water bodies. Sensibly, his supervisor, Mike Letnic, set up the study in the arid Victoria River catchment, where any toad that can't find water is in big trouble. Despite the toad's leathery skin, water evaporates from a Cane Toad at the same rate as it does from an open dish of water. As a result, toads have to replenish their water levels every few nights. Dan erected shade-cloth fences around dams in the late dry season, when temperatures are high and rainfall is nonexistent. He fitted radio transmitters to sixty toads and released them outside the fences. They were all dead within seventy hours, shriveled corpses beside the fence line. They had died trying to get back to the water.

Although keeping toads away from water can be devastating in the right place and time, though, that's not the case for much of the toad's Australian range. In the wet season, or in well-watered habitats closer to the coast, a toad has easy access to water. Even a mound of moist cow poo will suffice. Shutting off access to its favorite dam may disappoint the toad, but it won't be fatal.

Hand collecting, traps, and fences aren't the only approaches tried by toad-busting groups. Spraying juvenile toads with poisons was also popular, because it's impossible to pick up thousands of tiny scampering toadlets, but eventually it was banned by environmental authorities. When Graeme advocated firearms, the KTB responded that shooting toads was ineffective and dangerous.

Perhaps, rather than worrying about the mechanics of toad destruction, we should try to motivate toad collectors. A common theme of conversations in the pub is that toads could be controlled if there were money in it. Bounty hunters would scour the country, cleaning up every last toad. We could make a fortune by sending toads to China for use in traditional medicine. Sadly, China already has toads of its own.

How about the market in Japan, where toad toxin is used as an aphrodisiac and a hair restorer? I have bad news on that front. A decade of close contact with Cane Toads hasn't improved either my sex appeal or my hairline. I admit that completely useless products can sell in their millions (just look at bottled water). But are there enough lovesick, optimistic, bald Japanese men to sustain a commercial trade in toad toxins? And again, it would be cheaper to harvest feral Cane Toads on Ishigaki Island, off Okinawa.

We also hear about using toad products in expensive leather products, but it's a niche market—and a small one at that. The price of toadskin shoes is scandalously high, but only a few animals are used. I own a toadskin wallet, but I've never met anyone else who does.

Can we make money, instead, out of all those decomposing toad bodies? Frogwatch NT bottled decomposed toad gunk ("Toadjus") as a garden fertilizer, but the bottles had a regrettable tendency to explode. Customers ended up covered in gooey fermented toad. Even with Graeme's slick salesmanship, the product had limited commercial appeal.

At a broader level, providing a financial incentive to eradicate pests simply doesn't work. Commercial hunting encourages people to maintain stocks, not drive the quarry to extinction. I once heard a tragic conference talk by some researchers who discovered that a remote island had recently been invaded by a foreign frog. The frogs were only in one swamp, so the team paid the local kids to catch them. Many were brought in, but by the time the researchers returned to the island the following year, the frogs occurred in several other swamps as well. They had been spread by the young capitalists, who were keen to make more money when those generous foreigners came back.

A few less dramatic motivators for toad control were tried as the toad army swept westward, but they soon failed. The most popular promotion with my students occurred when the toads arrived in Darwin in 2006. A local hotel offered a free beer for every Cane Toad brought in. It was a marketing triumph for the groups that offered it, and the story went viral on the Internet. Predictably, though, the arrangement didn't affect toad numbers, and the offer didn't last for long. It's a thirsty climate, and there are a lot of toads out there.

Has collecting adult Cane Toads (by hand, traps, and fences) ever managed to eliminate toad populations? People make that claim, but it's a triumph of optimism over evidence. If you're anywhere within the toad's main range, it's impossible to see what effects you're having because the toads arrive from next door faster than you can get rid of them. The best opportunity to examine the effect of physically removing toads comes from small, isolated populations, founded by stowaways that came in on the back of a truck. In some of those cases, the toads have eventually disappeared. For example, Cane Toads arrived at the outback towns of Cunnamulla and St. George in western Queensland in the 1970s, but they were gone within two decades. They were brought in intentionally, to control insects in local gardens (extraordinary folly). But the toad's extinction was due to the harsh environment, not toad-busting. A small population a long way from the species' main range is unlikely to persist, because conditions are unsuitable. And even if the toads are released in the right place, an introduction at the wrong time of year may fail. That's what happened when toads were released south of Darwin in June 1967. The area is suitable for them—it's now awash with toads—but that first (mind-bogglingly stupid) introduction failed because it happened in the dry season. In other cases where toad populations have blinked out—such as near Port Macquarie on the New South Wales coast—we don't know whether the decline was due to toad-busting efforts or climatic variation. And we never will.

One thing we *do* know is that Cane Toads don't give up without a fight. Our efforts to eradicate Cane Toads from an industrial suburb in Sydney are

Cane Toads bred successfully in this small drainage pond behind a factory in the Sydney suburb of Taren Point. Photo by Matt Greenlees.

a good example. Hundreds of Cane Toads are accidentally brought to Sydney in trucks every year, but they rarely survive for long. Breeding is even less likely; because the eggs are fertilized after they leave the female's body, there's no such thing as a pregnant Cane Toad. It takes at least two toads, in the same place at the same time, to create a population explosion. Unfortunately, when stowaway toads arrived in a trucking depot near Sydney Airport in 2009, they managed to breed. Before long, residents of the suburb reported dozens of toads. Sutherland Shire Council sprang into action, but even with modern technology (like radio-tracking adult toads to find their breeding ponds), vigorous effort (regular patrols to find and remove toads), and the local rats (who eat the aliens), it took several years to eradicate the amphibians. We've collected more than seven hundred toads, but just when we think we've got them all, they breed in another drainage canal behind another industrial estate. We haven't seen a Cane Toad at Taren Point for the past three years, but an occasional one may still be emerging at night to feast on cockroaches. And even if we catch that last toad, another bunch of hopeful immigrants will be jumping off the back of a truck next year.

If you want to exterminate Cane Toads from an area, you need to think long and hard about two challenges. One is the proportion of the local toad population that you can catch. The other is how quickly any survivors can repopulate the area by breeding. The news is bad on both fronts.

First, how much does intensive collecting affect the abundance of toads? To answer that question, you need to know how many toads were there before you got to work, and how many are still there afterward. Reliable data on toad abundance are hard to come by, and the best information comes from Greg Brown's surveys at Fogg Dam. Greg found an average of 18.5 adult toads along the 1.5-kilometer (1-mile) dam wall in December 2008, a few years after the toads arrived. But that number dropped back to 6.4 per trip the following year, and was down to 1.4 by 2012. And this was in a place where nobody was removing toads. With this kind of background variation, it's difficult to measure the impact of control efforts on toad abundance.

The community toad-busting groups keep records of their capture rates, but counting casualties of the toad-bust doesn't get us anywhere. What we need to know is how those captures affect toad abundance. For that, someone has to go out *before* the toad-bust and mark and release every toad they see. The mark must be invisible to the toad-busters—say, a paint mark on the toad's belly. Then, when the toad-busters' haul comes in, you can see how many of the marked toads were caught. If all of them turn up in the toad-busters' bags, you can break open a beer—you may have caught all of the local toads. But if only half of them turn up, you've got more work to do. When Matt Greenlees followed volunteer toad-busters around in northeastern New South Wales in 2014, the volunteers failed to see most of the toads that Matt could see. That's not surprising. Many of the volunteers had never tried anything like this before. And even Matt must have missed quite a few. On a dry night, most Cane Toads stay in their burrows.

Of course, there are exceptions to every rule; and there are times and places where it makes good sense to collect adult toads. The best opportunity to eradicate toads is on small islands, where immigration of new toads is rare and there are limited breeding sites. Even here, though, major efforts have failed to get rid of toads. The owners of Hamilton Island in northern Queensland are putting in a lot of effort, but the toads are hanging on.

The only well-documented case where people exterminated a Cane Toad population was on tiny Nonsuch Island in Castle Harbour, Bermuda, only 6.5 hectares (16 acres) in extent. Cane Toads arrived there in 1978. The wildlife authorities set out to eradicate them. The only source of freshwater (and thus the only place toads could breed) was one small pond. Enormous effort went into fencing the pond to keep toads out, and to collect them both inside and outside the fence. But over the years the fence was occasionally damaged by wind, and the Cane Toads bred. It took five years, $10,000, and hundreds of person-hours of volunteer work to wipe out that population. That effort simply isn't feasible on a larger island or (even worse) an entire continent.

So, what can we say overall about the valiant attempts of Australian volunteers to control Cane Toad numbers by physical collection? Has it all been a waste of time? The answer depends on your perspective. It didn't slow the invasion; Cane Toads swept through the heavily toad-busted areas around Darwin and Kununurra at the same rate they swept through the "un-busted" Kakadu National Park a few years earlier. Intensive toad-busting certainly reduces local abundances of toads, at least in the short term. During the dry season, when toads aren't moving around, that impact can last for months. But it's not having any effect at the landscape level. Australia is a big country. Even in Darwin, some of Team Bufo's monthly sampling sites are the same as those regularly "toad-busted" by local volunteers. It doesn't prevent us from getting as many toads as we want, whenever we go there.

Toad-busting has a local short-term effect, but that's all. Whether or not that improves things for the local wildlife is difficult to tell. Most of the impact of Cane Toads occurs as soon as the aliens arrive, when they poison native predators, and it only takes a few toads to get rid of all the local goannas. Reducing toad abundance in already-colonized areas may confer very little benefit to biodiversity. Discouragingly, for example, the two-year period of sharp decline in toad abundance at Fogg Dam wasn't accompanied by any increase in the numbers of native wildlife. As yet, there is no evidence that toad-busting has helped the Australian ecosystem.

Long before my own involvement, the challenge of controlling alien Cane Toads was taken on by the Commonwealth Scientific and Industrial Research Organisation. CSIRO is a large, federally funded entity charged with conducting research in the public interest—but the definition of that interest has changed considerably over the past few decades. In the 1970s, when I was a student, CSIRO included a well-credentialed Division of Wildlife Research containing truly eminent ecologists. They conducted basic research to solve "applied" problems facing the Australian environment and agriculture. Many of CSIRO's well-funded, long-term projects led to breakthroughs with immense financial and ecological benefits. University-based scientists with heavy teaching loads looked with envy at their CSIRO colleagues in research-only positions.

As society changed, however, pressures on CSIRO increased. Funding cuts eviscerated the wildlife division. In 1989, though, a savior appeared. The federal government offered substantial money for research into controlling the Cane Toad, a politically attractive environmental enemy. A senior CSIRO scientist got a surprise phone call one afternoon, about the prime minister's forthcoming policy statement on environmental issues. What high-profile ecological catastrophe can we throw some money at? The CSIRO scientist said the first thing that came to mind: "Cane Toads."

And so, when Prime Minister Bob Hawke stood beside the Murray River the next day to deliver what he modestly called "the world's greatest environmental statement," he solemnly promised government funds to fight the plague of toads. Unwilling to look a gift horse in the mouth, the CSIRO team used the cash for field surveys—mostly in Venezuela, within the toad's native range—to look for viruses or other diseases that might be useful in controlling toads.

The decision to head to Venezuela was puzzling. If you want to find a virus to kill an invading species, the worst place to look is within the invader's native range. Any virus that occurs there has been infecting toads for thousands of years, and the toads will have evolved resistance to it.

A better approach was available, but the CSIRO team ignored it. One of CSIRO's greatest scientists, Frank Fenner, suggested looking in other toad species. A virus that makes another toad species ill might be devastating for Cane Toads. That approach had been the key to Frank's successful introduction of a virus that has killed untold millions of feral Rabbits. Originally found in a South American rabbit species, the disease (myxomatosis) was fatal to the European species that had been introduced to Australia—and was dramatically effective for controlling them.

The beauty of this approach—using viruses from the invader's close relatives—is that it is highly targeted. Diseases and their hosts are locked in an evolutionary arms race, with each adapting to the other. The virus keeps evolving to be more effective at surviving inside the host, and the host keeps evolving to combat every new trick that the virus comes up with. The virus ends up intricately adapted to its host and will make the host sick but not kill it. Viruses that kill their hosts won't be passed on; the best tactic is to infect a host and stay inside it, keeping the host healthy so that it pumps out new viruses for as long as possible.

The end result, then, is that most viruses (and bacteria, and other parasites) are very good at infecting their usual host species and don't cause too many problems for the host. But they can be devastating if they infect a related species that has never encountered them before. The virus can infect the new host, because it is not too different from the old one—but the new host hasn't had millions of years to adapt to the virus's "weapons" (ways to escape the immune response, and so forth). That process may blindside the new host's system, leaving it wide open for attack. Humans have suffered from such host-switching events in recent years, with the HIV virus (AIDS) moving across from Chimpanzees, severe acute respiratory syndrome (SARS) virus from chickens, and Hendra virus from Fruit-bats.

And it might have worked with Cane Toads. It's likely that a toad somewhere—perhaps *Bufo horribilis* (the "Cane Toad" species from the other side of the Andes) or some other species of toad in South America or Asia or Africa—has a virus that makes their normal host sick, but that would be deadly to Cane Toads in Australia. This is a well-known principle; indeed, Rob Slade and Craig Moritz pointed out this opportunity for toad control when they first described genetic variation within native-range Cane Toads in 1998.

The CSIRO Venezuela project ignored that option and instead searched for viruses within Venezuela. None of them proved useful. Either they didn't kill toads or else they killed every amphibian they infected—including Australian frogs as well as Cane Toads. So, the search for a native-range virus that kills Cane Toads (and only Cane Toads) was a failure.

What can you do if the natural viruses don't work? Make your own instead! As the government money continued to flow, CSIRO tried to genetically engineer a virus to kill Cane Toads. This, too, was an ill-advised approach. Even if you could overcome the huge technical challenges and produce such a virus, releasing it into the environment would rouse "green" activists to apoplexy. Indeed, it would be illegal in many Australian states. So, even if you constructed the perfect weapon, you could never use it. And, last but not least, suppose you produce the ultimate toad-killer: a virus that is fatal for toads but not for native frogs. How could you stop it from spreading to Asia, where there are many (perfectly innocuous) native toad species? Or even spreading to Brazil on some traveler's boots? Exterminating Cane Toads within their native range would be an environmental catastrophe worse than anything that the species has caused in Australia. After $12 million had been spent on the search for a toad-killing virus, the project was abandoned.

What about Frank Fenner's idea of looking for viruses in other species of toads? It's still a good idea, but it's too late. The futile search for virus-based control of Cane Toads chewed up too much cash and credibility. Our political masters are not inclined to throw good money after bad. A few years ago, Alex Hyatt and I suggested a (cheap) project to look for toad-killing viruses in other species of toads. Our plan was to send live Aussie toads off to various parts of the world, where they would be kept with native toads of various species. If the Aussies all dropped dead, we would know that the local toads contained a virus capable of killing Cane Toads. The answer to our funding request was a polite "No thanks." Viral control will probably never be applied to Cane Toads, but perhaps that's a good thing. We are playing a high-stakes game when we play God by moving organisms around the world. If we did bring in a virus, we'd have to be certain that it was harmless to Australian species before we released it. Once you let it go, it's unstoppable.

Basic research was the fundamental underpinning of the "old" way that CSIRO worked (back in its golden era). The first step in solving a problem was to understand it, even if that took a long time. If CSIRO had been asked to look into Cane Toads any time from 1935 to 1985, they would have set up an excellent research program. But that never happened. Cane Toads didn't become a political hot potato in Australia until the late 1980s, when they threatened to reach Kakadu National Park, a beloved national icon. By then, the CSIRO wildlife research team was already a pale shadow of its former self. If that ecological research on toad invasion had been done by CSIRO, I would have kept on studying snakes until I retired, because there wouldn't have been any point in duplicating the CSIRO effort. Team Bufo would never have existed. Such are the accidents of history.

By the time I began to think about Cane Toad control in 2006, nobody else was doing research on the topic. The CSIRO genetic-engineering work was going nowhere and would soon be abandoned. Could ecological research by Team Bufo find a different kind of weapon to combat the impact of invasive toads?

My first foray into toad control happened serendipitously. To understand the ecology of Cane Toads in Australia, I needed to know about their parasites. Medicine and farming tell us that parasites are incredibly important, sometimes making a host population crash rather than flourish. What parasites are found in Australian Cane Toads, and what effects do they have? And did those parasites accompany the toads from Hawaii or opportunistically switch across from Australian frogs?

I wasn't the first person to ask those questions. In the 1990s, Di Barton investigated the Cane Toad's parasites for her Ph.D. thesis at James Cook University. The most common parasite that Di encountered was a nematode worm that lives inside the toad's lungs; some toads had scores of these worms. The lungworms are small—less than a centimeter (half an inch) long, and as thin as cotton thread—but they might nonetheless be important, because they feed by sucking blood out of the fine vessels that line the toad's lung. Each worm is a hermaphrodite—it starts life as a male and produces sperm, but then undergoes a transgender shift to femaleness and produces eggs. This kind of sex change isn't uncommon in the animal world—for example, the Northern Territory's favorite fish, the Barramundi, undergoes the same midlife switch. Unlike the Barramundi, however, the worm stores those sperm that it made during its male phase, using them to fertilize the eggs that it creates later on. This lifestyle takes self-love to a whole new level.

The adult worm sitting inside the toad's lungs releases her fertilized eggs so that they can be coughed up into the toad's digestive tract and, eventually, be passed in the toad's feces. Tiny larvae that hatch from those eggs and live in the soil have conventional sex roles; the females mate with the males and produce their own larvae. These are "born" by bursting out of their mother's body (not a nice way to thank mum for all she's done for you, but parasites play by their own rules). The larvae then wander away from their mother's exploded carcass and look for a new home. They penetrate the first toad they encounter—usually by crawling across the top of its eyeball, into its brain, and then burrowing through its body to find the lungs.

Di's pioneering studies turned up an intriguing result: many of her lungworm-infected toads were sick. Were lungworms the cause? Maybe. Infected toads had lower blood cell counts and hemoglobin concentrations. Perhaps that was because of the worms sucking away at blood inside their host's lungs? If so, we might be able to use these lungworms to control Cane Toad populations.

Where did the toad's lungworms come from? When Di dissected native Aussie frogs, she found lungworms that looked the same as those in Cane Toads. If they were the same species, that situation could have happened in only two ways. Either the toads had brought American lungworms with

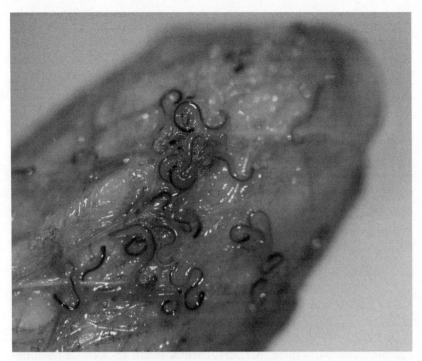

Lungworms were brought to Australia inside the founding Cane Toads, and they now prosper in toads over most of their Australian range. This photo by Crystal Kelehear shows a toad lung turned inside-out over a finger, revealing the lungworms embedded in the lining of the lung.

them and they had transferred across to frogs, or the lungworms from Australian frogs had transferred across to toads. The lungworms were common in frogs all over Australia, including places without Cane Toads, so the first alternative was impossible. This left the second one. Di decided that the lungworms from Aussie frogs had switched across to infect toads as well.

That idea made sense, because Cane Toads left many of their native parasites behind during the long trip from French Guiana to Puerto Rico to Hawaii to Queensland. With only a few individuals involved in each step, the founding fathers and mothers surely lacked some of the parasites that infected their source populations. And even if the founding adults contained parasites, the next generation of toads couldn't become infected if they were kept apart from their parents. That's what should have happened in Australia. Unfortunately, the sugarcane scientists didn't bother with quarantine precautions, so we couldn't be sure.

> **A PAUCITY OF PARASITOLOGISTS**
>
> Why do long delays keep creeping into these tales of parasite biology? Somebody does some research, and then there's no further investigation for a couple of decades. The reason is depressing: Australia has very few scientists with the specialized skills needed to identify parasites. And it's getting worse as the older generation retires. Few bright young researchers are coming up to replace them. The average parasitologist I meet is seventy-five years old, with a hip replacement—but is so passionate about his or her work that he or she is still limping into the lab every Wednesday to look at miniature worms through the microscope. Society's neglect of fields like parasitology (and taxonomy in general) will one day come back to bite us in the bum.

The story sat there for a long time—until 2006, when Team Bufo expanded into parasitology. Crystal Kelehear began research on the lungworm that Di had identified from Queensland Cane Toads twenty years earlier. Crystal grew up on a cattle farm, and she has a farmer's work ethic. When I had the temerity to suggest that she didn't need to gather so much data, I was greeted with a look of withering scorn. So what if it would take another month of peering down the microscope? Not a problem. Easier than getting up at 4 A.M. every morning to castrate bulls.

Crystal's experiments confirmed Di's suspicion: these lungworms are bad news for their hosts. If you infect a baby toad, it's likely to die; if it does survive, it will be small and slow. At its full size of about 8 millimeters long (almost half an inch), the parasite is more than half of a young toad's body length. For a human being, it would be like having several half-grown pythons inside your lungs, each about a meter long and as thick as your wrist. So that's why the infected toads aren't happy. And the larvae that grew into those adult parasites reached your lungs by burrowing through your brain. Small wonder, then, that lungworms can ruin a toad's day.

Could we use the lungworms to control Cane Toads? First, we need to be sure what species we are dealing with. If (as Di believed) the toad's lungworm were an Australian species, then it might be a useful biocontrol. If the native frogs already have the lungworm, we're not spreading a new parasite around.

There could still be problems—for example, the toads might provide many new hosts for the lungworms, building up their numbers enough to increase infection rates in frogs as well—but this would eventually happen regardless of what we did. All in all, a native lungworm that killed Cane Toads would be a useful ally.

On the other hand, the lungworm might still be a foreigner, an American species transported with the original toads from Hawaii. Given the poor quarantine in 1935, a parasite could have made its way to the cane fields. Because many lungworm species look similar, the best way to tell them apart is by looking at their DNA. Fortunately, my research group in 2007 included Sylvain Dubey, a dapper young Swiss biologist who specializes in this kind of study. I expected the parasites in Australian Cane Toads to be the same species as in Australian frogs. A boring "tick the box" project, but important to nail down before we could move on to the interesting stuff.

Two weeks later, Sylvain burst into my office with exciting news. The lungworm inside Australian Cane Toads was genetically identical to lungworms from South American Cane Toads—and very different from the lungworms found in Aussie frogs! So, Cane Toads had brought the parasites with them from South America.

This raised a problem for the idea of using lungworms to control toads. If the American lungworms can also infect Australian frogs and have nasty effects on them (as they do on toads), it would be idiotic to use lungworms for toad control. Not quite as stupid as importing Cane Toads into Australia in 1935, but close. We need to stop spreading ecological devastation around the world.

To find out if the toad's lungworm infects native frogs, we dissected frogs as well as toads. We collected the carcasses of road-killed frogs before the next truck obliterated them, dissected their lungs, and then obtained DNA sequence data on all the lungworms we could find. The news was good. Sylvain's analyses in 2007, and follow-up work by Ligia Pizzatto a few years later, showed that the only lungworms in frogs were native Australian species, and the only lungworms in Cane Toads were the American species. Apparently, the evolutionary distance between frogs and toads was too great for them to share parasites.

But hang on. There's another (and more sinister) possible explanation for why road-killed frogs lack toad lungworms. Infected frogs may die before they can hop across the road in front of a truck. To test that possibility, Ligia exposed young frogs to the toad lungworm in the laboratory. She found that toad lungworms could infect frogs (bad news), but the frog's immune system was able to deal with them (good news). An American lungworm that ended up inside a frog was unable to find the lungs and was killed by the host's immune system.

But, in terms of toad control, what's the point of infecting Cane Toads with lungworms that they already carry? The answer is that not all Cane Toad populations carry the lungworm. At the fast-moving invasion front, densities of toads are low—and, thus, parasite larvae deposited in toad feces are unlikely to find a new host. The consequent lack of parasites at the toad front may explain why toads thrive for those first few years after they colonize an area, until their parasites catch up with them. We could curtail that honeymoon period by bringing in lungworms.

Before we break out the champagne and welcome the lungworm as Our Great Hope for Toad Control, however, we need to step back a bit. Australia contains over three hundred different species of frogs. American lungworms couldn't survive inside the first few frog species that Ligia tested, but other species of Aussie frogs might be vulnerable. So Ligia plowed on with her trials. Raising hundreds of tiny baby frogs takes enormous time and effort, and was possible only because of funding from the Western Australian Department of Parks and Wildlife.

The results, when they came, were very disappointing.

The first hint of trouble came in 2011, when Ligia discovered that the American parasites could reach the lungs and survive inside one species of native frog. The infected frogs pooped out hundreds of lungworm larvae, just like infected toads do. And, to make it worse, the vulnerable frog was my favorite amphibian: the big, squat-bodied Green Tree Frog. Fortunately, though, the frogs weren't fazed by the little bloodsuckers. Although the parasites thrived in the frog's lungs, they didn't reduce their host's survival, growth, or reproduction. Relief all around! I wondered if we could use Green Tree Frogs to spread the toad parasite around, like a fat green Typhoid Mary, carrying the parasites but not affected by them, able to spread them willy-nilly in the face of the toad invasion.

But then it all came crashing down. The Green Tree Frog's ability to host these parasites made us nervous. Would other frog species not only take up the parasites, but also be killed by them? Ligia decided to look at the Magnificent Tree Frog, which, as its name suggests, is an awesome animal. Frogs don't get called "Magnificent" very often, but this one deserves it. It's big and green—like the Green Tree Frog—but it has white spots as well, and large glands that give its head a comical appearance. These frogs are charming, and that's not a word I'd apply to most amphibians. Magnificent Tree Frogs are found only in the Kimberley region of Western Australia, right where the toads were heading, so Ligia wanted to know what would happen when these amphibians met the American lungworms. The answer was depressing. Magnificent Tree Frogs took up the parasites but couldn't tolerate them. Every infected frog was dead within a few weeks.

Because Magnificent Tree Frogs live mostly in rocky gorges and well off the ground, they aren't likely to encounter too many toad parasites—but still, it means that (at least in the Kimberley) we can't spread the American parasites around to control toads. If we do, we'll kill native frogs as well.

Crystal came back to Team Bufo to do her Ph.D. work from 2008 to 2012, and her follow-up studies revealed another problem for using lungworms to control toads at the invasion front. Larval lungworms are tiny and can't survive for long in the harsh outside world. The larval parasite dies unless it rapidly finds a new host. As a result, a lungworm population can't persist if its hosts are rare—that's why lungworms are absent from the toad invasion front. Indeed, that pressure is so great that lungworms near the toad invasion front have evolved larger eggs, which give the young worms enough energy to keep them alive until they can find a new toad. If we spread lungworms around near the toad invasion front, where toads are scarce, most of the parasites will die before they find a host.

Another nail in the coffin came when Greg looked at how lungworms affect the rate of dispersal of their warty hosts. Based on simple logic ("bloodsucking parasites make you weak and tired"), I expected the parasite to slow toads down. If so, lungworms might reduce the rate at which the toad invasion front moved into new territory. Greg checked the idea, in 2014, by raising baby toads to adulthood, then infecting half of them, and radio-tracking both infected and noninfected animals. The parasites didn't affect toad movements. I was ready to give up on lungworms for toad control, until an honors project by Patt Finnerty in 2016 tried a different way to measure the impacts of these parasites on toads—and got a different result. Patt injected free-ranging toads with a drug that kills lungworms and found spectacular differences between dewormed and "control" toads. Infected toads didn't move or feed as much, and they were well-and-truly sick. In laboratory tests, for example, the infected toads were slow and weak, and they struggled less than usual when picked up.

Patt's comparisons also revealed some intriguingly specific effects of the parasite on its toad host. Lungworm infection caused toads to select hotter refuge sites by day, to poo in water rather than on dry land, and to produce copious, watery feces—and all those changes pay off for the lungworms. A hotter toad produces more lungworm larvae in its poo, and a lungworm larva that is deposited in watery feces (especially in a moist site) has a much higher chance of survival than if it was deposited in a drier location. So Patt's data suggest that lungworms are manipulating the behavior of their hosts, probably by sending specific chemical messages. Research on other species is revealing similar stories; the effects of parasite infection aren't just to make hosts ill. Instead, parasites have evolved ways to manipulate their hosts into becoming better vehicles for spreading the parasites' progeny to new hosts.

Will the lungworm offer a useful component of toad control? We still don't know. It certainly won't be useful in the Kimberley, where Magnificent Tree Frogs would be imperiled. But, given the profound effects that Patt found, perhaps we can still think about using lungworms in other areas, such as the southern invasion front in New South Wales. But only after we check out the vulnerability of the local native frogs! Our lungworm studies are a good example of a new idea arising about how to control an invasive species—and how further research shows that it won't work, or will work only in certain conditions. That's the fate of most attempts at biological control; 99 percent of the brilliant new ideas crash and burn.

So, was the parasite work a failure? Not at all. We learned how an invasive species and its parasites interact with native fauna. The Australian Research Council funded the work because of its scientific value, not for pest control. Understanding host–parasite interactions will help us manage future invasions, of all kinds. But although the idea of using parasites for toad control in the Kimberley was dead and buried as far as I was concerned, there was a hiccup. One of the toad-busting groups—the KTB—had heard about Crystal's earlier work on parasites and decided that it was the perfect answer for toad control. And, not trusting the Sydney scientists, the KTB set up their own research program.

The KTB heard about lungworms at a talk I gave in Perth in April 2007, when I first suggested that the American parasites might help us control Cane Toads. It was just an idea, and we needed more work to see if it was feasible. That work knocked the idea on the head (at least for the Kimberley), for reasons I've explained above.

But the KTB were excited about the lungworms. Full steam ahead! Unwilling to rely on Sydney-based academics, the KTB began their own lungworm research in 2008. Their active Internet presence soon attracted a potential researcher. Many European students come to Australia to conduct short-term projects. They get credit for these studies in their home institutions, and a chance to travel and enjoy a warmer climate. A Dutch student found the KTB online and offered to join them. I suspect that he got a few surprises when he got off the plane in Kununurra, but it was the beginning of a long relationship. Jordy Groffen is a very tall, intense young man with a background in agricultural studies. A search of his name on Google turns up Jordy's master's thesis from 2012 (at Wageningen University in the Netherlands) entitled "Tail posture and motion as a possible indicator of emotional state in pigs." This may not be the *ideal* background for a toad researcher; then again, I was a snake biologist before I began working on Cane Toads.

Jordy traveled around with KTB teams, dissecting toads and running experiments to infect them. And when a Cane Toad symposium was held in Darwin in 2008, with community groups as well as scientists invited, Jordy gave a talk about his lungworm research. It was obvious to the scientists in the audience that there were problems with Jordy's experiments, but the questioning after his talk was polite. Nobody wanted to criticize a young man, probably talking at his first conference, and in a foreign language (a terrifying prospect). And Jordy had had very little contact with professional scientists, because the KTB discouraged interactions with the enemy.

Things unraveled in the question-and-answer session after the talk. A Queensland University professor asked Jordy to tell us a bit more about the life cycle of the parasite. It was a well-intentioned question, designed to fill in the embarrassing pause when nobody can think of anything to ask the speaker. Unfortunately, Jordy had no idea about several very basic issues. The questioner then asked me if I could describe the lungworm's biology, and I deferred to Greg (because he knows much more than I do about these parasites). Greg's explanation of the life cycle of rhabditoid nematodes, gently correcting mistakes in Jordy's presentation, was as polite as he could make it. But we all cringed at how awful it must have been for Jordy.

I went up to Jordy at the next tea break to make friendly small talk. That wasn't easy for either of us. It wasn't just the social tension, but also the fact that Jordy is six-foot-seven, whereas people often refer to me as "the short bald guy." Standing side by side, we were in a long-distance relationship—and we had to shout to be heard. At the time, I attributed his shell-shocked look to being out of his depth (like an amateur football player trying out with the pros). But maybe it was just the light reflecting off the top of my head. Whatever the reason, Jordy left the conference feeling that professional scientists were his enemies. He never accepted my invitation to visit Middle Point. He was probably reflecting that research on the emotional meaning of piglets' tail movements would have been a better long-term career choice after all.

Not long afterward, the newspapers, radio stations, and TV shows trumpeted a major research breakthrough. Jordy had found lungworms in toad populations only a year behind the invasion front, whereas "other scientists" mistakenly believed that it was decades before the worms caught up. The KTB put out excited press releases about Jordy's "ground-breaking discovery." When asked by journalists, I declined to comment. I never doubted that Jordy had found lungworms close to the toad front line, but I didn't understand why that was important. And why were the lungworms turning up at the toad invasion front only in the areas where the KTB was toad-busting?

We now know how those lungworms reached the toad invasion front: the toad-busters transferred them as they moved from one site to the next. In 2011, Crystal showed that it was a simple issue of contamination. Lungworm larvae can live in damp mud inside the tread in a boot sole for at least a week, and toad-busters going from one area to another had carried the parasite around with them. It was an interesting pattern, but Jordy didn't realize that it was due to human activities and not to parasite biology.

However, Jordy's Great Lungworm Discovery—that lungworms sometimes occur in toads close to the invasion front—showcased a less attractive side of the KTB: their view of toad-busting as a political battle. Their press releases proudly proclaimed that "the scientists" had been wrong—we (that is, I) had erroneously thought that the lungworm front was at least twenty years behind the toad front. Where they got that idea, I don't know. By the time Jordy was hitting the newspapers with his story, we knew that the lungworms were generally about two years behind the toad front—but the KTB hadn't read our papers.

One of the many interviews that Jordy gave about his research breakthrough was in a Dutch magazine on April 18, 2008. The text reads as follows (translated from the Dutch):

> What Australian researchers have been looking for for years, the Dutchman Jordy Groffen (20) from the city of Haarlem has found in two and a half months of his internship in the bush. He found a solution to eradicate the plague of Cane Toads. Australian scientists—who received hundreds of thousands of dollars in research subsidy—are looking like an embarrassment.... "Whilst dissecting the Cane Toads I found the worm in their lungs," says Groffen. "Small toads die from these lung worms. Adult speci-

mens aren't able to grow as big and have a hard time getting around." It turned out to be the solution. Groffen catches the infected animals and puts them out in a healthy population of Cane Toads. "They infect each other and it slowly eradicates them. Very simple." Thanks to his discovery, Groffen is big news in Australia. Only scientists are less pleased. According to him, the well-known toad professor Rick Shine tried to steal the research of Groffen. "He asked me if I had infected toads for him, he found my research interesting and wanted to run with it himself," according to Groffen, who quickly put his findings on the Internet. "It is my discovery."

If I were a defamation lawyer, I would have been licking my lips. Instead, I sent Jordy a polite "please explain" email. In his reply, Jordy apologetically acknowledged that the idea of using lungworms for toad control came from Team Bufo, long before he arrived in Australia. And that bit about me wanting to steal his research? A complete fiction. The journalist had made it up.

I'm skeptical. Another KTB employee made the same accusation—that I had stolen Jordy's research—to one of my students a few years later. It's strange how this bizarre story emerged twice from within the same small group. But perhaps that blistering Kimberley sunshine boils the brain and encourages unusual ways of thinking. Or could it be yet another unexpected impact of Cane Toads? If people handle too many toads, mysterious chemicals seep into their skin and rewire their brains? That might explain Reg Mungomery's observation about Cane Toads being the epicenter of a world of nonsense and fabrication. Damn toads! I knew it was their fault somehow!

NINE

A New Toolkit for Fighting the Toad

All war is deception.

SUN TZU, *The Art of War*

Suppose an alien army is spreading through tropical Australia. It's easy to find and kill a few soldiers, but the enemy sends more. One way to fight that invasion is to send troops out into the field, to kill the bad guys (the toad-buster approach). Another is to develop high-technology weapons of mass destruction (CSIRO's "genetically engineered virus" approach). But there's a third way—gather some intelligence. Analyze what the enemy is doing, what resources he relies on, how he communicates. Armed with that knowledge, you can launch devastating counterattacks that hit him where he is weakest.

Military strategists would support the third of those tactics. Cracking the Nazi code system (a breakthrough built on basic scientific research) did more to win World War II than any battleship. But "finding out more about toads" was unpopular with the people advocating direct action against the horrible tide of croaking death. When asked to fund research on Cane Toads, one government minister declared that "we've spent enough bloody money on scientists measuring toads—let's just work out how to kill the bastards!"

I understand why people feel this way. "Scientific research" often looks like an excuse for nerds to satisfy their own curiosity. A focus on practical outcomes looks to be a better investment. That works well for small, easy-to-solve issues. But history tells us that complex problems are solved by "basic" rather than "applied" approaches. Most major medical and technological advances were spin-offs from basic research—often into topics that, at first sight, had little to do with the problem in question.

The ill-fated CSIRO toad project was my first introduction to Cane Toad research. In 2002, I was drafted onto a committee that oversaw the program in its final years. Some good research had been done, but we were still a long way from understanding toad impact, let alone developing methods for control. I'm not competent to evaluate CSIRO's work on viruses, but it did develop a method to detect Chytrid Fungus, a major threat to native frogs. The most interesting insights into toad biology had come from Ross Alford at James Cook University and from his Ph.D. students Mark Hearnden and Michael Crossland (later to join Team Bufo).

By the time the toad invasion front crossed the Queensland border into the Northern Territory in 1983, the toads were moving faster but research had slowed down. In this remote area, fieldwork was expensive and there were few people to complain about the invading amphibians—and thus less political pressure to stop them.

Although my initial focus was on basic biology rather than pest control—to understand toads rather than annihilate them—I occasionally dreamed of discovering a "magic bullet." And as our data accumulated, that idea evolved from a forlorn hope into a distinct possibility. I changed my mind about Cane Toads. Rather than seeing them as invulnerable, unstoppable invaders, I came to see them as animals out of place. Cane Toads were encountering problems as well as causing them. Perhaps we could exploit those weaknesses to achieve (dare I say it) toad control.

Our first attempt to exploit an amphibian Achilles' heel centered on the lungworm parasites that toads had brought with them from Hawaii. As I described earlier, that project was a scientific triumph but its value for pest control remains unclear. In the process of deciphering the relationship between the parasite and its amphibian hosts, we discovered that spreading lungworms widely across the Kimberley to inconvenience Cane Toads would be a death sentence for native frogs.

The second option for toad control was a simpler idea. It emerged from research into the sex lives of Cane Toads, so I'll digress here to talk about that topic.

As in most amphibian species, female Cane Toads grow bigger than males. The evolutionary reason is simple. A bigger female has more space to carry her eggs, which enables her to produce more offspring than a smaller toad. For males, size isn't so important. Even a small male produces enough sperm to fertilize a vast number of eggs.

Male and female Cane Toads differ in other ways as well. For example, male toads have rough, sandpaper-like skin, whereas females are smooth.

Females remain brown throughout their lives, whereas a male toad turns yellow when he is reproductively active. In my opinion, a bright-yellow male toad is a good-looking amphibian.

Male and female toads differ in their lifestyles, too. The males hang out around the breeding pond while the females forage in the surrounding countryside. Female toads are gluttons, because it takes a lot of energy to produce a clutch of eggs. And when a female is ready to breed, romance is easy to find. The males gather in groups around a suitable pond, pouring their heart and soul into a long guttural call that has been likened to the noise made by a diesel engine, a motorcycle, or a machine gun. The calls of Cane Toads have found their way into the soundtracks of movies like *Star Wars* and *Jaws*.

Attracting females is tiring, because yelling nonstop for hours really chews up energy. The upside is that toad calls are loud—an animal the size of your fist can be heard a kilometer (half a mile) away. Every female in town knows where you are, and what you have in mind for her. And you are caught in a dilemma. If you find a solitary calling site, any female you attract will be yours and yours alone. If instead you join a chorus with other males, you can create a wall of sound that is more likely to lure a receptive female to your pond—but then you'll have to fight for her favors against a dozen rivals.

The fierce rivalry among males explains other sex differences in Cane Toads as well. In the breeding season, male toads change shape: the muscles in their front legs swell up like those of weight lifters, and they develop thickened pads on their front feet. The male hops onto his lady friend from above and behind, embraces her by slipping his arms down around her sides and into her armpits, and then hangs on grimly. That position (known as "amplexus") places his rear end close to hers so that when she extrudes eggs, he can release sperm to fertilize them. His rippling muscles and roughened footpads help him cling to the female when other males try to take his place.

With the edge of the pond swarming with males, mistakes are inevitable. A hyped-up male toad jumps on anything the right size, including plastic bottles and human feet. Amorous males grab other males so often that they have evolved special "release calls" to tell their overexcited rival that he's made an error. A male toad gives this distinctive chirping call as soon as he is grasped, even if it's a human doing the grabbing. A female toad can't give that call because she doesn't have vocal chords. As a result, a male Cane Toad has a simple rule. If you grab something that doesn't give a release call, hang on for grim life. This creates problems for any toad-shaped object (such as a native frog) that happens to be in the wrong place at the wrong time. Even if

A pair of Cane Toads in amplexus, with the male firmly holding his partner. Photo by Crystal Kelehear.

that frog is a male, frantically giving the correct release call for its own species, the sex-crazed toad on top of it won't believe what it's saying. Because frogs and toads don't recognize each other's language, many frogs drown in the embrace of love-struck toads. In 1871, the extraordinary ardor of male toads moved Charles Darwin to write, "Although cold-blooded, their passions are strong."

Around the edge of almost every breeding pond, there are many more male toads than females. A female only comes down to the pond once or twice a year, after she has accumulated enough energy to produce a clutch of eggs, whereas male toads are ready for sex every night. Any female that approaches the margins of the pond encounters a throng of males intent on grabbing her first and asking questions later. The female is soon the focus of a wrestling match, with the original male holding on for dear life while new arrivals try to pry him loose. That's why male toads develop forearm muscles like bodybuilders.

But finding the right partner isn't just about brute force. An intriguing experiment on European toads showed that the males judge their opponent's size not by how big he looks, but by how big he sounds. Bigger male toads have deeper croaks, just as bigger men have deeper voices. In deciding whether to hang on to a female or relinquish her to a competitor, a male toad listens to his rival to judge how big he is. The experimenters tested this idea by putting rubber bands around the males' throats so they couldn't croak, and playing tape-recorded calls that ranged from squeaks (from little males) to deep croaks (from the biggest males). Sure enough, a big male would give up and let go of the female if he heard a deep croak during the battle—even if his rival was a pipsqueak.

One of my students, Haley Bowcock, followed up this British study with experiments on Cane Toad sexuality at Middle Point in 2006, by sitting out at night beside toad-filled enclosures until the wee hours. When I arrived back from fieldwork, I'd see the silhouette of a tall woman crouched beside her toad tubs, in the faint glow of a dim red light, surrounded by a million swirling mosquitoes. In those trials, Haley showed that lugging a male around is a problem for a female toad. After he grabs hold of her in amplexus, he rides piggyback on her for hours or even days. She has to carry him around wherever she goes, making it difficult for her to move around and feed. But it's not easy for her to get rid of him, because she can't give the "release call."

Does this mean that a female Cane Toad is vulnerable to any male she encounters? No. It's more complicated than that. The males call loudly, swagger around, try to have sex with anything that moves, and generally make a nuisance of themselves. The females are more demure. So, when biologists first started looking into amphibian sex, their studies were focused on male behavior. A female was just an object for the males to fight over.

But we now understand that in many species, females run the show. For example, a female toad that is grabbed by a male she doesn't like can swim over to a male she fancies and thus precipitate a fierce combat for her favors. In some elegant studies at Middle Point, Bas Bruning and Ben Phillips showed that females have another trick as well. Toads can inflate themselves with air, swelling up like a balloon to frighten off predatory snakes. Female Cane Toads use the same tactic to discourage an unsuitable lover. Even when he's got a firm grip behind her armpits, it's difficult for the male to hang on when his female friend swells up to twice her size. So, females have more say in mating outcomes than we'd think at first sight. Cane Toads will never be a symbol for the women's liberation movement, but there's more sexual equality in the frog pond than meets the eye.

When a female Cane Toad ventures into motherhood, her output is spectacular. The eggs are tiny black objects, less than 2 millimeters (a tenth of an inch) in diameter, like a row of black pearls embedded in a transparent string of jelly. The eggs are fertilized after they leave the female's body, as the male squirts sperm into the water. Most often, the long egg-strings are wound around grass stems in shallow water.

I've never had the self-discipline to count how many eggs are present in a single clutch. That's what students are for. Their counts range up to 40,000 eggs per clutch; the clutch comprises about one-third of a female's body mass. High fecundity is common among amphibians, but Cane Toads take it to an extreme. A large toad has room to carry a lot of small eggs.

The tadpoles that hatch a few days later are miniature also; I'm amazed that all the organs of a vertebrate body can fit inside such a tiny creature. But they soon start swimming around and feeding. They use hard grinding teeth, on the upper and lower jaws, to rasp away at algae on the bottom of the pool or the scum floating on the pond surface. Those mouthparts are the only parts of the animal that aren't gooey.

Tadpoles are creatures of mystery, and there is much that we don't know about their lives. For example, why are they fond of each other's company? Not only do toad tadpoles prefer some places to others (they especially like sun-drenched shallows), but they also hang around in schools. These can be loose, with lots of stragglers, or so tightly packed they look like a black football. Milly Raven discovered that the cues for aggregation are social. Cane Toad tadpoles keep a close eye on each other and are strongly attracted to the sight of a feeding fellow tadpole. So, as soon as one tadpole finds food, it is surrounded by a school of its fellows.

Adult Cane Toads are slothful, spending the day in a deep trance like an amphibian Buddha. But their children are the opposite. Cane Toad tadpoles zip around in perpetual motion, as if they had Amphibian Attention Deficit Disorder. Like all children, they focus on eating, pooping, and growing.

Within a few weeks (or months, if food supplies have been poor), the tadpoles grow legs, resorb their tails, and turn into toadlets. We are so familiar with this transition—metamorphosis—that we no longer wonder at it. But it's incredible. A tadpole is a small, fish-like creature that eats algae, breathes through gills, and is mostly a small sphere of intestines propelled by a large tail. That polliwog transforms into a miniature toad that eats insects, breathes through lungs, and hops about on four legs. To my mind, it's as impressive as the evolution of fishes into frogs—and that took a few million years, not a couple of days.

Continuing the theme of miniature size, juvenile toads are less than 10 millimeters long and weigh less than one-tenth of a gram (one three-hundredth of an ounce). Having developed the basic body structure (and ecology) of a land-dweller, these little guys are in for one hell of an adventure. If they survive, they can grow to more than a kilogram (over 2 pounds). Their shape doesn't change much, but they develop from a tiddler into a monster.

It's worth reflecting on that span. By contrast, a baby human (around 3.5 kilograms, or 8 pounds) increases only about twenty-fold in weight to get to its maximum size (say, around 70 or 80 kilograms, or 165 pounds). A better comparison is with our size when we start independent life (say, at around eighteen years of age) compared to our final size. Baby toads are independent from the day the egg is laid, so that's the critical step from an ecological perspective. Humans barely change in size from independence to adulthood, whereas toads show a 3,000-fold increase.

During your own life, you've changed in size and ability, had different jobs, and so forth, but the toad makes you look like a stick-in-the-mud. At metamorphosis, a Cane Toad is small enough to be eaten by the insects that will eventually be its prey. The native frog that it meets by the edge of the pond will be a dangerous predator when the toad is a baby, a competitor for food when the toad is a juvenile, and a snack for the toad when it's an adult. Simple classifications of another species as "predator," "competitor," or "prey" don't work.

If our young toad finishes her tadpole stage in the dry season, she emerges into a harsh and dangerous world. The only place she can survive is the moist mud around the water's edge, where she has to compete for food with thousands of her brothers and sisters. If she is one of the first out of the pond, though, her prospects are better. She has access to more food, so she can grow rapidly before the main onslaught of hungry mouths. Her increasing size lets her forage farther from the pond's edge, in drier areas where there are more bugs to eat. And, if she's lucky, she reaches a size where she can fit a smaller toad into her mouth. From then on, she's living in a supermarket.

Around a pond, the young toads fall into two size groups. Among the thousands of tiny black ones, not long out of the tadpole stage, there are a few dozen bigger toads. Brown and splotchy in color, like an adult toad, these are the cannibals. They grab a bug if it wanders past, but they prefer bigger game. And when a toad turns to the dark side, it switches foraging tactics. It sits still and wiggles its toes in rhythmic waves that lure a hungry smaller toad in search of a wriggly insect. The stomachs of these larger juveniles are crammed with micro-toads. And this rich food source enables them to grow even faster, outstripping their unfortunate younger relatives even more.

The only way for a tiny toad to escape the Cousin From Hell is by changing its time of activity. Older toads—including the cannibals—move around only after the sun sets. A young toad that feeds by day, and goes to bed early, can avoid becoming supper for a larger toad. Cannibal feasts take place at dusk, when young toads that have tarried too long at feeding encounter their larger relatives.

Fewer than one egg in a thousand survives long enough to produce an adult toad. Most of the mortality occurs early in life—in the egg, tadpole, and metamorph stages. Any young amphibian that survives through to the next wet season has a far rosier future. The onset of regular rains liberates the survivors to wander off to greener pastures, where the youngsters can focus on feeding and growing.

How can an understanding of Cane Toad sex suggest new ways to control these invaders? As we all know from movies and novels, episodes of passion create risky situations. And that vulnerability is especially great if our hero or heroine is in a strange country, surrounded by dangers that they never encountered back home. In Australia, Cane Toads are in exactly that situation.

Adult Cane Toads are ecological generalists, willing to disperse into any terrestrial habitat that offers food and water. That makes them a formidable enemy, because their troops are scattered over the entire landscape. They would be far more vulnerable if they relied upon just a few, very specific kinds of sites. We could identify those sites in advance and annihilate the toads as soon as they arrived. And exactly that opportunity is offered by the Cane Toad's sex life. Every adult toad must return to a water body if it is to produce offspring. Even better, a critical phase of the life history—from eggs, through tadpoles, to metamorphs—is restricted to those spawning ponds and their margins. That dependence on rare, predictable sites is a key vulnerability of the invading amphibian, one we can use to inflict maximum damage.

And there's a second vulnerability: Cane Toads evolved in South and Central America, where climatic conditions, plants, and other animals are very different from those in Australia. On one hand, it's astonishing that a thin-skinned rainforest amphibian from Latin America has thrived in the driest continent in the world. On the other hand, there are ways in which

Cane Toads are poorly suited to conditions in Australia. In the jargon of wildlife biology, mismatches between an invader and its new home are "ecological traps": a way of behaving that evolved in one situation can lead you into deep peril in another. If you didn't evolve here, you'll act inappropriately in some situations, or fail to handle some new challenge. If we can identify those mismatches, we can exploit them to control the invader.

Australia has been isolated from the rest of the world for a very long time; it's been more than 50 million years since Cane Toads and Australian frogs last shared a common ancestor. The differences that have arisen over that period make it easier for us to develop control methods that affect the invader but not the natives. That's a critical issue. It wouldn't be difficult to kill most of the Cane Toads in Australia: we could launch nuclear strikes on every billabong. Unfortunately, though, we'd kill everything else as well. The challenge is to eradicate toads without harming the native wildlife. And to do that, we need to find unique features of the invader. We need to find "ecological traps" that render the invasive amphibian vulnerable without imperiling the native fauna.

So, what mistakes are Cane Toads making in Australia—especially with respect to where, when, and how they breed? The first "ecological trap" we found involved the rules that Cane Toads use for selecting a breeding site. That discovery was made in 2005 by my student Mattias Hagman, a frog enthusiast with a shaved head, rippling muscles, and plentiful tattoos (reflecting a former life as a drummer in a Swedish rock band). Mattias drove around Kakadu National Park to look at where toads were breeding—and, just as importantly, where they weren't. Within a week of beginning his study, Mattias identified strong patterns in the attributes of toad spawning ponds.

First, Cane Toads don't like flowing water, so streams are out. And all of the ponds where Mattias found breeding toads were shallow, with open and not-too-steep muddy banks. Most were roadside verges and construction sites that fill up with water during wet-season rains but are dry the rest of the year. Native frogs prefer deeper and more densely vegetated breeding ponds. In any given area, Cane Toads breed in only a few ponds—generally the ugliest water bodies in the landscape. Two years later, in 2007, Mark Semeniuk repeated Mattias's study at the other end of the Cane Toad's Australian range, in the eucalypt woodlands of New South Wales. Mark found that southern toads love the same kinds of trashy ponds as their tropical relatives.

Why do Cane Toads prefer these water bodies? By choosing open, muddy banks, male toads ensure that their advertisement calls are not muffled by vegetation. And female toads prefer shallow water and gradually sloping banks, where they won't drown when a sex-crazed male (or a duo or trio) grabs them. To really understand those patterns, though, we'd need to study the breeding biology of Cane Toads in their native range in Latin America. Over millions of years, Cane Toads that bred in shallow, open water bodies must have been more successful than those that bred elsewhere. Those habitat preferences became hardwired into their DNA and are still exhibited by their Australian descendants.

Understanding the toads' choosiness has an immediate benefit: we can predict where toads will breed and then target our control measures at those sites, knowing that few native species will be present, thus minimizing collateral damage. We don't have to waste our time checking out unsuitable ponds. And it's easy to change a pond to make it less of a Toad Magnet. For example, some councils now insist that property developers grow dense vegetation around any ponds they create, to remove Bufonid Boudoirs.

The seasonal timing of reproduction is another "ecological trap" for Cane Toads. In the Australian wet–dry tropics, it's hot, dry, and dusty for nine months of the year—and hot, wet, and luxuriant for the other three months. Native frogs reproduce at the onset of the monsoons, around Christmas time. A spectacular amphibian orgy erupts with the first drenching rains (frogs really know how to party). But Cane Toads ignore the opportunity. After the rains stop in March or April, the hot, dry days return with a vengeance. And it's then, early in the dry season, that the Cane Toads of Fogg Dam ramp up their sex lives.

Why do Cane Toads delay, rather than exploiting the monsoonal paradise? There are two reasons. One is the nature of potential breeding sites. When the rain stops, high temperatures cause the temporary ponds to shrink day by day, leaving a widening ring of mud around the water bodies—an ideal platform for a male toad's nocturnal serenades. The second reason involves energy. Unlike native frogs, Cane Toads remain active year-round—and, as a result, they lose body condition during the long dry season, when food is scarce. When the first rains arrive, a female Cane Toad doesn't have enough energy reserves to produce eggs. It takes a few months munching on insects in the lush wet-season pastures before she is fat enough to reproduce.

In their native American range, where the climate is less seasonal, Cane Toads reproduce all year. Laying your eggs as water levels drop isn't a problem, because it will rain again soon. But the equation is much tougher in the Australian wet–dry tropics: many ponds dry out before the toad tadpoles can

Carnivorous ants kill vast numbers of metamorph Cane Toads. Photo by Greg Clarke.

complete development, dooming the entire clutch. And there's another disadvantage to dry-season sex as well: insect predators like water beetles have been breeding throughout the wet season, and those drying pools contain vast numbers of hungry bugs: sometimes, enough to slaughter every toad tadpole. The seasonal "ecological trap" enables managers to concentrate their control efforts in time as well as space, and to benefit from the volunteer army of native predators.

Those "ecological traps" extend to the Cane Toad's interactions with the native fauna in other ways as well. In 2008, Georgia Ward-Fear showed that large carnivorous ants eat enormous numbers of young Cane Toads but take very few native frogs. The Aussies know that ants are dangerous, and they move away from them. In contrast, a toad ignores the ant until it's too late.

In 2011, Elisa Cabrera-Guzmán found a similar story with water beetles and water bugs. Native frog tadpoles are adept at evading these fearsome miniature monsters, so the water bugs dine on toad tadpoles instead. To encourage these voracious insects to eat even more toads, we could simply clean up our water bodies so that they support diverse insect communities.

Another "ecological trap" involves competition: tadpoles are each other's worst enemies. A small pond doesn't contain much food. More tadpoles mean that everyone grows more slowly, and the weaker ones die. It's a zero-sum game. It doesn't matter if those extra mouths belong to other Cane Toad tadpoles or to other species. Australian frog tadpoles are bigger than toad tadpoles—sometimes tenfold heavier. These big bruisers eat most of the food. As a result, adding frog tadpoles to a pond is a disaster for the Cane Toad tadpoles that live there—creating a terrific opportunity for low-risk, environmentally friendly toad control. Over much of Australia, native frogs have been wiped out by pollution, disease, and domestic cats. We can reverse that decline. Many of the places where Cane Toads breed are close to towns, easy for us to get to. And if we clean up the area, we'll make it worse for toads (but nicer for frogs) anyway. Even better, the most effective toad-suppressor is the Green Tree Frog, an Australian icon. Their huge tadpoles strike fear in the hearts of tiny toads-to-be. These huge, dumpy, bright-green frogs, with perpetual goofy grins and big toe-pads on their stout little feet, are happy to live close to people.

During my childhood, Green Tree Frogs were common in the Sydney suburbs. More than once, I felt sticky little toes on my hand when I reached inside the letter box for mail. Environmental degradation has driven Green Tree Frogs out of the cities, although they are still common in bushland areas. Is it feasible for us to bring them back, as heroic Green Avengers to take on the toads? Yes, but it will require a rethink by wildlife managers (see sidebar: "The Ethics of Keeping Wildlife as Pets").

What else can we do to control Cane Toads? Surely research can generate solutions a bit more high-tech than growing pondside vegetation and releasing native frogs? In 2007, an unlikely set of experiments finally gave us the silver bullet that everyone had hoped for. The hero of the story is Michael Crossland. He's an introverted, unflamboyant hero—a Clark Kent rather than a Superman. But it was Michael's ideas that ultimately bore fruit—after many twists and turns, and a few blind alleys along the way.

The breakthrough centered on an idea that sounds preposterous, at least initially: *We can control Cane Toads by exploiting the ways that they interact with each other.* But if you give it some more thought, there is logic to the notion. All animals interact with members of their own species, and some of those interactions have sinister functions. They enable one individual to thrive at the expense of another. If Cane Toads have evolved a way to outcompete other Cane Toads, we might be able to exploit it. For example, if they use chemical warfare to kill other toads, those chemicals could give us a new arsenal of anti-toad weapons.

THE ETHICS OF KEEPING WILDLIFE AS PETS

Native frogs can be powerful allies in our fight against Cane Toads: frog tadpoles eat the same food as toad tadpoles and can outcompete them. A pond full of frog tadpoles is a hostile environment for a toad tadpole. So, should we breed up frogs in captivity and then release them into the wild, in order to control Cane Toads? I think we should, but that isn't as easy as it would have been thirty years ago. Attitudes about keeping wildlife in captivity have changed. Frogs and reptiles weren't legally protected when I was a schoolboy. I could go out and catch a lizard and bring it home to keep in a cage in my backyard. Nowadays, native frogs and reptiles are officially protected, and you need a mountain of paperwork to catch an animal and bring it out of the wild. Kids and paperwork don't mix, so people keep cats, dogs, and budgies rather than frogs and lizards. And the idea of keeping wild animals in captivity has become politically incorrect.

Some animals aren't suited for cages. The sight of a Sulphur-crested Cockatoo in a birdcage always saddens me, because they are too smart to thrive in a small prison. But frogs and lizards aren't Einsteins, and many of them are happy in a safe cage, with a comfortable shelter and plentiful food. It's a nastier life out there in the bush than many people imagine: predators, competitors, and times when there's no food or it's too hot, too cold, too dry, or too wet. Parasites and diseases aren't much fun either.

Opposition to keeping native animals in cages has driven creatures like Green Tree Frogs out of our homes. But what a perfect animal to keep in a well-maintained aquarium in the classroom or living room; and how exciting for children to see the miracle of metamorphosis—from a fish-like brown tadpole into a green froglet. And what great motivation for community groups to rehabilitate the local pond. Pull out the old shopping carts, restore the pondside vegetation, and release a thousand tadpoles produced by your captive Green Tree Frogs. If we can give the next generation hands-on experiences with native animals, those children will grow up to be eco-warriors. The word *wildlife* will evoke thoughts of real-live Blue-tongue Lizards and Green Tree Frogs, not digital images of elephants and lions on the TV screen. And if we can bring native frogs back into our suburbs, the world will be a more difficult place for Cane Toads. Occasionally a clammy little foot will grab our hands as we check the mail, but that's a small price to pay.

But that would require Cane Toads to compete with each other intensely. Is that a reasonable scenario? Yes, it is. Everywhere you find them (in Brazil as well as Queensland), Cane Toads live in human-degraded habitats. Other species are scarce in such places. If you find a Cane Toad as you wander around at night, the next animal you'll find will probably be another Cane Toad. As a result, Cane Toads mostly compete with each other, not with native species. Adult toads competing for bugs under a streetlight are the most obvious example of rivalry, but competition for food is even more intense in the murky waters of the local pond. Tadpoles can't move away when the going gets tough. If you can wipe out your competitors, your own chances of survival improve considerably.

Those ideas suggest that we should be looking at tadpoles, and at the ways they communicate with each other. The first aspect of tadpole "language" that we looked at in 2006 is a signal that says, "Watch out, there's a predator nearby." Birds have special alarm calls for that function, and a single *"cheep"* can send an entire flock hurtling into the air. Remarkably, the tadpoles of several toad species produce an antipredator signal also. It doesn't rely on sound, because tadpoles don't have vocal chords or ears. In dirty water, vision is useless as well. So tadpoles rely on chemical signals to interact with their fellow polliwogs.

When a toad tadpole is injured, it releases "alarm chemicals" into the water. The chemicals are stored inside goblet-shaped cells (named *Riesenzellen*, surely a useful word for Scrabble) that can be emptied rapidly if the tadpole is stressed. A toad tadpole that detects the chemical knows that a fellow tadpole is under attack, and it might be next. As soon as it encounters that chemical, any sensible tadpole does a U-turn and looks for somewhere to hide. In technical parlance, the alarm chemical is a "pheromone"—a substance that is secreted by an animal and influences the behavior of other members of the same species.

This is weird. A toad tadpole is tiny—less than 10 millimeters (not even half an inch) long—and has a brain the size of a pinhead. How can a little lump of gunk have a sophisticated early-warning communication system like this? But it does. Mattias Hagman and I ran many trials, crushing a toad tadpole between our fingers (an instant and painless death) and dribbling the resultant fluid into an experimental tank of water. Any toad tadpole instantly turned tail and fled as soon as it got a whiff of the alarm chemical. That presumably benefits the receiver of the signal—it can escape from danger—but what happens if you bombard a tadpole with such signals? Is it stressful to encounter the alarm pheromone on a frequent basis? Prolonged stress can be deadly; is that the case for the alarm chemical also?

Yes. In an experiment that we ran in outdoor ponds at Middle Point in 2008, toad tadpoles that encountered the alarm pheromone every day were unlikely to survive. Constantly encountering a message saying "danger, danger, danger!!!" was too much for them. Even if they survived larval life, the pheromone-exposed tadpoles turned into miniature toadlets that were highly vulnerable to other risks (like drying out, or being eaten by ants or by other toads).

Before we could use the alarm chemical for toad control, however, we had to be sure that it doesn't also terrify the tadpoles of native frogs. Fortunately, it doesn't; the tadpoles of native frogs don't even seem to detect it. The next hurdle was bigger: it isn't feasible to go around crushing millions of toad tadpoles to generate the material for our scare campaign. We needed to identify the critical chemicals involved. To do this, we collaborated with an eminent chemistry professor at the University of Queensland. Rob Capon has a laboratory full of gleaming machines that can take a drop of fluid and work out every chemical within it. He was keen to help us find that magical toad-frightener. As a Queensland resident, he probably lies awake at night, feeling guilty about the initial introduction of Cane Toads in 1935. Or at least he should.

My naive early idea was to take some juice from a crushed tadpole and separate out a few chemicals and test them. I remember the look of pity in Rob's eyes as I explained my foolishly optimistic hopes. I hadn't appreciated how many thousands of different chemicals make up a toad tadpole! But Rob's background in chemistry told him that some substances were more likely to have the "alarm" effect than others. Rob selected some promising candidates, and we ran lab trials to see if any of them scared the bejesus out of toad tadpoles. Michael conducted thousands of those trials at Middle Point and identified a few chemicals that made toad tadpoles run the other way. So far, so good. But when we ran those same chemicals in longer-term trials, to try to reduce tadpole survival and growth, nothing worked.

What's going on? Toad tadpoles are smarter than I thought. When a predator (like a fish or larval dragonfly) grabs a toad tadpole and injures it, a cocktail of chemicals is released. And toad tadpoles respond to all of those substances. If bits of your brother float past, it's not worth hanging around until you detect some specific body part. Instead, just get the hell out of there. If you keep detecting just a single chemical, it probably means that nobody is actually massacring your siblings. We're continuing those trials, but I'm not optimistic. Even if we could find the magic "alarm chemical," it's unlikely to be useful for controlling toads. It doesn't reduce tadpole survival dramatically enough, and we'd have to find a way to keep adding it to natural ponds for weeks at a time.

Although the alarm chemicals weren't useful for toad control, they encouraged us to work with experts in chemistry. That collaboration paid off when the real breakthrough came along, involving a different response. Like the alarm pheromone, the chemical involved is produced by toad tadpoles—but it causes other larvae to rush toward it, rather than to flee.

Our discovery of the "attractant pheromone" is a good example of how answers emerge in unlikely ways. Michael and I were studying the ecological impact of Cane Toads, not looking for ways to control them. When toads spawned in small ponds near Fogg Dam in 2010, the bodies of hundreds of dead native-frog tadpoles floated to the surface and lay in piles at the water's edge. Lab trials confirmed that frog tadpoles try to eat Cane Toad eggs but can't tolerate the poison. The tadpole is dead within a few minutes. It was a depressing result. Despite decades of discussion about the ecological impact of Cane Toads in Australia, nobody had ever noticed the most numerous victims of the toad invasion: the tadpoles of Australian frogs. There was no doubt that in the ponds around Fogg Dam, thousands of frog tadpoles were being fatally poisoned by eating toad eggs. But did this carnage have any effect at the population level? The species most at risk would be those whose tadpoles actively search out Cane Toad eggs, rather than just eating any that they happen to encounter. A tendency to actively search for toad eggs would be a death sentence for a tadpole—and perhaps for a species.

To see whether some kinds of frog species were at more risk than others, we put out funnel-traps (like the ones used to catch small fish) in the ponds where we had found the spawning toads and the dead frog tadpoles. Some of our traps contained toad eggs, whereas others contained "control" bait (dog food, which all tadpoles like). If tadpoles of some species of frogs actively search for toad eggs, we should catch lots of them in the egg-baited traps. But in a species whose tadpoles just eat toad eggs when they stumble across them, the numbers of tadpoles in "control" traps should be as high as in "toad egg" traps.

I left the field station and went back to Sydney that weekend, before we could finish the trials. Two days later, I got an excited phone call from

Michael. The traps baited with dog food had caught a few dozen tadpoles, of several species. No surprises there. But the traps baited with toad eggs had caught vastly more. When Michael went out to check, his toad-egg traps were surrounded by thousands of frantic tadpoles pushing each other out of the way to get into the trap. Incredibly, though, they weren't the frog tadpoles we had been expecting. Instead, they were Cane Toad tadpoles! Our experiment had answered the question we had originally asked: the tadpoles of native frogs eat toad eggs if they happen to find them, rather than actively searching for them. That was good news. Even better, we had accidentally discovered a way to catch vast numbers of toad tadpoles.

Why did Cane Toad tadpoles flock to the eggs of their own species? It was a search-and-destroy mission. As soon as a tadpole found an egg, it began rasping away at the jelly coat to reach and consume the embryo inside. Toads are immune to their own poison, so this wasn't a Larval Last Supper. A frog tadpole dies as soon as it eats a single Cane Toad egg, but a toad tadpole keeps on munching. If there are toad tadpoles in the pond already, many newly laid toad eggs become dinner for their older relatives.

At first sight, cannibalism like this (and among metamorphs around the edge of the pond) is a puzzle. Wildlife documentaries often tell us that animals do things that benefit other members of their species, but the sad truth is that nature doesn't work that way. The natural world is a tough place. Animals evolve to do things that enhance their own chances of surviving and reproducing. Helping out other members of your species usually isn't favored in the rough-and-tumble world of natural selection. Instead, it's all about an individual leaving more copies of its own genes in the next generation. After she spawns, a female toad travels back into the forest to feed, and it will take her many months to build up the energy for another clutch. By then, the offspring from her previous clutch will be toadlets, not tadpoles. Those new eggs, then, are not the brothers and sisters of the tadpoles already in the pond—they are enemies. If those eggs hatch, the older tadpoles will have to share the food in their pond with an extra 40,000 hungry mouths. Natural selection doesn't have ethics. Organisms evolve to do whatever pays off in terms of their own survival. In this case, obliterating the opposition is the best strategy. Even better, you get a free meal as well, and a chance to top up your own store of poison.

So that's why Cane Toad tadpoles are cannibals—but *how* do they find those newly laid eggs? The pond where Michael had conducted his trapping experiment was a farm dam, built to provide a drinking source for the

Water Buffalo in the paddock. One of the stranger personal habits of Water Buffaloes is their determination to defecate in water, not on dry land—and that pond was the communal latrine for a hundred buffaloes. As a result, the contents of the pond were more accurately described as "diluted buffalo poo" than as "water." It was impossible for the tadpoles in our traps to have seen the eggs until they bumped into them. They must have found them by smell—and sure enough, our laboratory trials confirmed that hypothesis. As soon as they detect the chemicals oozing out of eggs, toad tadpoles forget whatever else was on their tiny minds and head toward that alluring scent as fast as their muscular little tails can push them.

As soon as we discovered this response, we knew that we had a potential new method for controlling toads—by luring the tadpoles into scent-baited traps. Our lab trials early in 2011 confirmed that although the scent of toad eggs brings Cane Toad tadpoles running, it has no effect on the tadpoles of native frogs. We had set out to understand the impact of Cane Toads on native frogs, and in the process we had blundered across a new and powerful way to catch toad tadpoles. And (even better) our method didn't catch frog tadpoles.

This was only the first step, though. It wasn't feasible to obtain kilograms of toad eggs to make the "bait" for our traps. We needed to identify the chemicals in the toad's catnip. Fortunately, Rob Capon and his student Angela Salim were enthusiastic about a joint attack on the problem. They sent us samples of promising chemicals (with exotic names like "suberoyl arginine," "H-Leu-Leu-OH," and so on), and we tested them in lab trials at Middle Point in 2011. Before too long, we hit the jackpot—we identified the attractant molecule. And, with the benefit of 20/20 hindsight, I felt like an idiot for not having guessed the solution immediately. The chemical we were trying to identify had to be distinctive—something found in Cane Toads but not frogs. What kind of chemical is unique to Cane Toads? Ask anybody in the local pub, and their first guess would be the toad's toxin: the powerful poison that a Cane Toad uses to discourage its predators. Sure enough, one of those poisons—bufagenin—proved to be the key attractant.

I wasn't the only scientist to feel a trifle embarrassed. After we published our results in 2012, several colleagues sent me photographs they had taken of Cane Toad tadpoles feeding on dead adult toads, but—like me—they had never realized that this behavior could provide a new approach to toad control. Most embarrassing of all, a few years earlier I had asked Mattias to study the responses of toad tadpoles to toxin. I thought that the toxin might be the

Cane Toad tadpoles flocking into a toxin-baited trap. Photo by Michael Crossland.

"alarm pheromone." Mattias came into my office a few days later, chuckling—not only did toad tadpoles *not* run away from the toxin, they were actually attracted to it! But attractant pheromones weren't on my radar at the time, so I ignored the observation.

Identifying the role of toad toxins gave us a ready source of the tadpole-magnet chemical. Although adult toads put some of their poison into the eggs, they store most of it in their shoulder glands. If the bulging parotoid glands of an adult toad are gently squeezed, a white, sap-like poison oozes out. The squeezing has to be done carefully, because this stuff is deadly. If the poison squirts out into your eye, or a cut in your finger, you're in big trouble. It's a job that requires protective clothing.

Knowing that toad poison attracts toad tadpoles was an exciting first step, but it was still a long way from eradicating them from natural ponds. To find out if we could do that, we needed field trials. A young Japanese researcher, Takashi Haramura, helped Michael set out traps in ponds near Fogg Dam. We used simple funnel-traps (open-topped plastic boxes with inward-pointing funnels) baited with toad poison: one trap for every 10 meters (11 yards) of shoreline. The results exceeded my wildest expectations. Every trap caught thousands of Cane Toad tadpoles. In the first two ponds we trapped, we eradicated toad tadpoles in less than a week. We got about 4,000 tadpoles in that first farm dam, so we upscaled and tried a bigger one. That larger pond yielded 40,000 tadpoles. In both ponds, we caught about 90 percent of the tadpoles within the first three days. Michael and Takashi carefully checked the pond margins over the next few weeks for baby toadlets. There were none to be seen. Our traps had got them all. Even the usually reserved Michael couldn't keep the smile off his face.

This new weapon in the war against toads dovetailed perfectly with my emerging thinking about Cane Toad control. Simply removing adult Cane Toads can never reduce the species' abundance in the long term, because toads breed at a phenomenal rate. They can replace themselves faster than we can remove them. But if we can stop toads from breeding (for example, by eradicating tadpoles from local ponds), every toad we remove from the backyard is one less toad in the population. The ones left behind can't produce new toads to replace the ones that have just been popped into the freezer.

Michael didn't rest on his laurels. Even before we had published the paper describing the attractant pheromone, Michael discovered that Cane Toad tadpoles produce yet another kind of chemical—one that may be even more useful. I call it "Bufonite," from the toad's scientific name *(Bufo)* and the chemical that brought the comic-book superhero Superman to his knees ("kryptonite"). The name has been used before—in the Middle Ages, to describe the mythical gem thought to be present within the head of a toad. But *Bufonite* is a fine word and worth recycling. There'll never be a simple, single answer for toad control, but Bufonite may be our most powerful weapon yet.

The story begins in 2010, with some lateral thinking by Michael. If toad tadpoles are hell-bent on destroying newly laid eggs, cannibalism isn't the only way to do it. Perhaps they have also evolved chemical weapons? Rather than rely on "hunt 'em down and eat 'em," a toad tadpole might produce chemicals that make life difficult for new eggs. The eggs hatch within a couple of days, so there isn't much time for tadpoles to detect, locate, and attack a new clutch. Do tadpoles also produce a waterborne chemical that disrupts egg development? If so, without even needing to find the eggs, the "search and destroy" chemicals could prevent the young upstarts from turning into competitors.

Tadpoles murdering other tadpoles with killer chemicals? That sounds more like the plot of a science fiction novel rather than a research program.

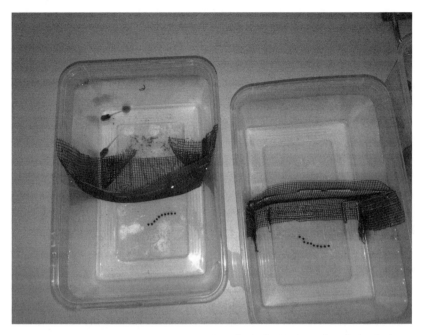

A simple experimental system to measure the effect of tadpole-produced chemicals on toad eggs. One set of eggs (left) is exposed to water containing tadpoles, while the other (the "control"; right) has no tadpoles on the other side of the mesh partition. Photo by Greg Clarke.

But tadpoles are sneaky. They engage in fierce warfare, using weapons that make a military arsenal look like child's play. And those weapons all target the same enemy: other tadpoles. For example, the feces of some tadpoles contain algae that kill younger tadpoles. Other larvae produce inhibitory chemicals, like the toad's alarm pheromone. Although nobody had ever found a chemical that affected amphibian eggs rather than tadpoles, the idea wasn't completely crazy.

Michael devised a simple, elegant way to test his idea. He placed a vertical partition (of fine wire mesh) across the middle of a small plastic box, dividing it into two equal halves. Then he poured water into the box. The water could flow freely through the holes in the mesh, but tadpoles couldn't fit through. Michael then put a few freshly laid toad eggs on one side of the partition. The other side either remained empty (in half the boxes) or contained a few toad tadpoles (in the remaining boxes). The partition stopped the tadpoles from eating the eggs, so the only way they could affect them was by releasing chemicals into the water. If the chemicals had an effect, eggs with tadpoles in the other half of their box shouldn't develop as well as those with just water behind the partition.

The experiments were a stunning success. In the "control" boxes, the eggs produced healthy toad tadpoles. But on the other side of the mesh screen, in the boxes with older toad tadpoles, things went badly. About 20 percent of the eggs didn't hatch, and even the survivors were in trouble. A day after hatching, the "treatment" tadpoles were smaller than their brothers and sisters from the control treatment. These little, loser larvae grew into fragile, tiny tadpoles that developed slowly. And the few that made it through to metamorphosis turned into miniature toadlets—unlikely to survive in the harsh world that confronts a young amphibian.

And it just keeps on getting better (or worse, if you're the tadpole involved). It doesn't matter if the toad eggs are just-laid or close to hatching. The effect is always strong. It doesn't matter if the tadpoles creating that killer chemical are young or old: they all produce it. Water that has contained toad tadpoles (even one tadpole in 500 liters of water!) is enough to affect the developing eggs. Whatever those tadpoles are secreting, it's powerful stuff. The young tadpoles that emerge from those doomed eggs show many abnormalities. For example, although they don't rival Great White Sharks in terms of dental equipment, tadpoles *do* have keratinized tubercles that they use to scrape algae into their mouths. And the "suppressed" tadpoles have malformed teeth, or none at all. These are seriously buggered-up tadpoles.

That's exciting news if you want to control Cane Toads; if we can identify the magic chemicals that are so deadly to toad eggs, we can hoist the Cane Toad with its own petard. It's a chance to deliver a highly targeted strike against the invaders. One great advantage of Bufonite is that a brief exposure is enough to disrupt toad development. If we use the alarm pheromone, in contrast, we have to add it continuously. With the attractant chemical, we need to set traps and empty them. But with Bufonite, we only need a slow-release system to keep concentrations of the chemical high enough to clobber any toad eggs that are laid in the pond. Even a cage full of live toad tadpoles in a pond might produce enough Bufonite to prevent anybody else from breeding there.

And there's another possible benefit. Imagine you're a female toad (no need to get down on all fours or eat beetles; this can be a thought experiment). You're full of eggs, ready to breed, and are attracted by the baritone call of a male toad at a nearby pond. When you arrive he seizes you in a tight embrace, and together you wander into the shallows. But wait. What's that chemical you can detect in the water? Damn, it's Bufonite. There are already toad tadpoles in the pond; maybe your new male friend had a dalliance a week ago? That substance spells doom for any eggs that you lay. If you go ahead with your midnight tryst, you will waste an entire year's reproduction. Instead you should say, "Hey, let's go somewhere where it's a bit less crowded." A sensible female amphibian will only lay her eggs in a pond without that whiff of Bufonite.

Does the story play out this way? We don't know (yet). But in several other frog species, females won't lay their eggs in a pond containing older tadpoles. If Cane Toads act in the same way, and if Bufonite is the cue they use, we have an even better toad-control weapon. We can deploy Bufonite to prevent toads spawning. If we concentrate their breeding in a few easily accessible ponds, it will be much easier to catch and remove adults (and any tadpoles from toads that escape us). Toad control would become feasible at a landscape scale.

What about collateral impact? If Bufonite affects frogs as well as toads, the chemical is useless for toad control. Fortunately, though, toads speak their own language: like the other pheromones that we have tested, Bufonite is highly targeted. When Greg Clarke exposed the eggs of native frogs to toad tadpoles, those eggs grew into perfectly healthy frogs. Bufonite is a disaster for toad eggs, but not for anything else. In many ways, Bufonite appears to be an almost perfect weapon: effective at tiny concentrations, with no impact on native species. Even better, the life-history stage that it clobbers—the eggs—is the easiest stage to find. Unlike adult toads, eggs are predictably concentrated in the same ponds, year after year. It's a pest-control officer's dream.

When I talk about Bufonite, one of the things that people often ask me (once I've calmed down and stopped screaming "Hallelujah") is how this kind of "toad versus toad" warfare evolved in the first place. Answering that question took years of work by the next generation of Team Bufo's tadpole specialists: Jayna DeVore, Simon Ducatez, and, of course, Michael Crossland. The deadly impact of Bufonite on toad embryos is intimately tied to the rapid evolutionary changes that have occurred in Cane Toads as they spread across Australia. Remarkably, Bufonite has relatively little effect on Cane Toad eggs from Queensland (or from Hawaii, the source of the Australian toad population). It's only toad eggs from the invasion front—

from the Northern Territory and Western Australia—that are sensitive to the killer chemical.

The main effect of Bufonite, we discovered, is to accelerate development of the embryo. The most vulnerable period for a developing egg occurs around the time it hatches. The jelly coat has dissolved, so it's defenseless; but it isn't yet able to swim away from danger. If an embryo approaching this stage detects the presence of older tadpoles in the pond—by encountering even a tiny amount of Bufonite—its best option is to avoid the risk of cannibalism by developing as fast as possible through the vulnerable stages. And that's what happens—toad embryos that detect Bufonite shift into top gear and develop through to the free-swimming tadpole stage much faster than usual. That speeded-up development moves them out of danger sooner than would otherwise have been the case.

But if the response to Bufonite increases the survival of embryos, why do so many of them die if they detect it? Because accelerating development uses up energy reserves that a young tadpole needs for other functions. It's no problem if the egg is large—the tadpole can afford to spend that extra energy on escaping from the phase of development when it is vulnerable to its cannibalistic older relatives. But if the egg is small, an embryo that tries to accelerate development runs out of energy and either dies or fails to grow into a functional tadpole.

In Queensland, Cane Toads produce relatively large eggs: competition among tadpoles is intense, so the young tadpole needs to start life with as much energy in reserve as possible. For those large eggs, the ability to respond to Bufonite is helpful—if an embryo detects that it is at risk of being gobbled up by an older tadpole, it accelerates development to rapidly reach the stage where it is safe. But Cane Toads produce much smaller eggs at the invasion front. For the past eighty years, individuals on the invasion front have lived in a world where bugs are plentiful and other toads are scarce. Competition from other toads is almost nonexistent, and there is little chance that older tadpoles will be present in a spawning pond. That shift in evolutionary pressures has resulted in a switch in life-history tactics. Faced with the same amount of total energy to devote to her clutch, a range-core female from Queensland divides it up into fewer, bigger babies while a range-edge female from Western Australia opts for numbers over size.

That decrease in egg size has evolved so quickly that the eggs of invasion-vanguard toads are caught in an "ecological trap." Their small size renders them vulnerable to Bufonite. When a tiny embryo tries to ramp up its developmental rate (having detected the presence of older tadpoles), it runs into trouble because it doesn't have enough energy reserved. Rapid evolution of egg size has created a novel vulnerability, one that we can exploit to control Cane Toads in the places where they cause the most ecological damage—at the invasion front.

TEN

Toad Control Moves from the Lab to the Field

In 2005, the biological effect of the cane toad was listed as a key threatening process under the Environment Protection and Biodiversity Conservation Act 1999.

AUSTRALIAN GOVERNMENT, 2011

There is a massive chasm between having a new idea and translating it into practice. The history of research is littered with brilliant ideas that never made it out of the laboratory. And when it came time to translate knowledge into outcomes, I felt like I was in kindergarten. My training and experience had taught me how to make discoveries—but in professional research, the end result is a paper in a scientific journal. Full stop. Implementing our findings would raise a host of new challenges and take me far out of my ivory tower.

Fortunately, though, I knew where to start. Three major results from Team Bufo's work were screaming out to be applied: our new ways to (1) control toad numbers (by preventing breeding, using tadpole pheromones); (2) buffer the impact of toad invasion at the front (by educating predators to avoid toads, using "teacher toads"); and (3) to prevent the further spread of toads (by exploiting an arcane mix of ideas about toad behavior, physiology, and evolution). I'll discuss each of these in turn.

The most straightforward outcome from our work was a new way to reduce the abundance of toads. Target tadpoles, not adults, using an integrated approach based on our newfound understanding of toad reproduction—and, especially, of the sneaky devices by which toads compete with each other. These ideas fit easily into mainstream thinking about pest control, and into the ways that people were already battling toads—solving an invasive-species problem by killing the invaders. Community groups had already shifted from one method to another—from hand collecting to traps and fences—as the futility of each technique became clear. The toad-busters would be happy to shift again if we could provide a new recipe for toad destruction.

Our earliest insights were about breeding-pond preferences and the virtues of native insects and frogs as toad-controllers. Straightforward, eco-friendly, easy for people to implement around the garden pond. When I was interviewed about our research on radio and television, I blitzed the airwaves with a simple message: Cane Toads are not invulnerable, and the general public can make life more difficult for the much-loathed invaders without having to thump toads with golf clubs. My wife, Terri, and I set up a website (http://www.canetoadsinoz.com) to preach the virtues of cleaning up suburban ponds. Cane Toads thrive in habitats that have been degraded by human activities, so restoring the pond environment makes it a better home for a frog and less suitable for a toad. It's not rocket science.

Getting publicity was easy, because our research attracted extensive media coverage. And we got better at it as time went on. For example, Georgia Ward-Fear's project about ants attacking metamorph toads went viral because she had video footage of ants dismembering toads—exactly the kind of thing that TV news programs like to insert for entertainment in the nightly broadcast. Predictably, the story was misunderstood in some quarters. One woman phoned me in tears, saying that her daughter was allergic to ant stings, and my plan to import millions of giant ants would be perilous for her. It took me thirty minutes to convince her that I had no such plans. One KTB stalwart told me that during fieldwork, his Aboriginal friends camped beside ant colonies because "that professor on the TV said that the ants would keep the Cane Toads away."

Michael's discovery of the "attractant pheromone" raised a difficult problem, however. We now had a way to eliminate Cane Toad tadpoles from water bodies, but there was a catch. The methods are simple, the equipment is cheap, and the "bait" is free—just squeeze the shoulder glands of a dead Cane Toad. Tadpole-trapping wannabes bash adult toads whenever they see one, providing a ready source of tadpole attractant. But did I really want to encourage the public to kill toads and extract their poison? A misdirected squirt of toxin could have devastating repercussions not only for the person who was splattered, but also (potentially) for the professor who had recommended the practice on the TV news.

Deluged with requests from the public about how to trap toads, I produced an illustrated guide to the method, then sent it to the university's lawyers. After a collective scream of horror, they agreed that I could release

HOW TO HUMANELY KILL A CANE TOAD

Cane Toads are hardy. In the suburbs of Darwin in June 2008, a Cane Toad swallowed by a dog was still alive when it was regurgitated forty minutes later. (The dog's owner then kept the toad as a pet, naming it "Spew.") Cane Toads can survive incredible injuries (like having their eyes pecked out by birds, or being badly burned by bushfires) and just keep on hopping. So killing a Cane Toad is a challenge. The toad isn't going to make it easy for you.

Researchers kill toads using specialized drugs that aren't available to the general public. Community groups gas Cane Toads with carbon dioxide, but that's not practical for most people. Vigilante toad-busters kill the alien amphibians in many different ways: by hitting them on the head, by spraying them with poison, by decapitation—and some people pop them in the freezer. Which of these methods is the most humane?

Can research answer that question? Yes, but there's a problem. If you tell an animal-ethics committee that you want to measure the amount of pain an animal feels as you kill it, they react as if you had asked them to parade naked through the streets singing Gregorian chants. I can't blame them. Imagine a headline in the morning paper saying "Boffins Torture Frogs to Find Out How to Cause Pain."

So, what is the most humane way to kill an adult Cane Toad? It's an important question not only for animal welfare, but also for the peace of mind of members of the public who need to kill toads. Physically battering a toad to death is a stupid idea. It is hazardous both for the toad (if the blow is misdirected) and the operator (a squirt of toxin in the eyes is excruciating). But if physical force isn't the answer for Toad Termination, what is?

How about the once popular method of putting the toad in a plastic bag inside the fridge for a few hours, then popping it into the freezer? Unlike warm-blooded animals, a toad doesn't fight cold by shivering to generate heat. It just slides into unconsciousness. The same thing happens every night in many places where toads live, so it is not stressful. As the temperature keeps dropping, the toad's nervous system and brain gradually shut down. In the freezer it turns into a toadsicle and never wakes up.

Unfortunately, "cooling then freezing" is illegal under animal-ethics regulations. It was outlawed in the 1980s, after an ethics committee argued that the outside of a toad (or frog, or fish) would cool faster than its insides, which might allow its fingers and toes to freeze while the brain was still warm enough to recognize pain. If so, the

> animal would suffer in those last few moments of its life. Although that report didn't provide any evidence (it was just an opinion about what *might* happen), everyone jumped on the bandwagon. All around the world, ethics guidelines banned "cooling then freezing" as a way to kill cold-blooded animals.
>
> In 2014, I decided to look into the "cooling then freezing" method myself. I was dubious about the ethics-committee argument. When a toad goes into the freezer, surely its brain hits subzero temperatures almost as quickly as its toes? And, importantly, cold is an anesthetic (that's why plastic surgeons use ice packs). I suspected that a Cane Toad's brain would stop working long before its toes froze.
>
> To test this idea, we needed to measure what's going on inside a toad's brain after we place it in the freezer. My student Josh Amiel took on the challenge. In collaboration with Adam Munn and John Lesku, I got ethics approval to implant electrodes into the brains of live toads. The toads wore the loggers around like top hats for a few days, unworried by their small helmets. A few days later, the toads went into the fridge and then the freezer. Just as we'd expected, their brains quietly and steadily shut down. No bursts of activity, as we'd see if they were suffering. By the time ice crystals formed in their toes, they were oblivious to everything.
>
> But the saga is a sad one. For decades, Cane Toads (and other small cold-blooded animals) have been killed in barbarous ways, because a simpler and kinder method of euthanasia was outlawed based on flimsy pseudo-facts. All of us—scientists, the animal-welfare movement, and animal-ethics regulators—owe an apology to a lot of animals that died in agony, for no good reason.

the document—but only after they peppered it with multiple disclaimers, in lurid red capitals, absolving the university of blame if things went wrong. With bright colors, vivid photographs, and huge capitalized warnings, the instruction booklet resembled a KTB newsletter.

As I had hoped, the toad-busting groups were overjoyed to have a new weapon that was easy to use and diabolically effective. They leapt on the technique like a dried-out Cane Toad hurling itself onto a damp cowpat. Within a few weeks, the newspapers were full of photographs of happy toad-busters posing beside mounds of dead toad tadpoles. Predictably, the new method was presented as a product of research by the toad-busters, not the boffins—but I was delighted to see Michael's discovery taken up so quickly.

That uptake has continued to expand. We've now trialed our trapping system from Cairns to Kununurra, and we've taught community groups how to do it as well. "Team Bufo Traps" are being deployed widely in New South Wales, Queensland, the Northern Territory, and Western Australia, with outstanding results. In the first year they tried, both Frogwatch NT and the KTB reported huge hauls (one of the few things they have ever agreed about). At the time of this writing, people using Michael's method have taken literally millions of Cane Toad tadpoles out of water bodies across Australia over the past year. For the first time, we are making a dent in Cane Toad recruitment.

But we still need to fine tune the system. For example, we need a chemical that lasts longer and isn't dangerous to children or domestic pets. Our recent trials have come up with some strong candidates, so I think we can solve that problem. To develop a toad-control package that can be sold commercially, we joined with our Queensland University collaborators in patenting the magic method. And we are working with a private pest-control company as well—innovative thinkers, with long experience in the practical issues that are so foreign to a researcher's world. Hopefully, the Team Bufo Trap will soon be on the shelves in hardware stores.

Frustratingly, our efforts to determine the chemical identity of Bufonite (the suppression pheromone) are still based in the lab and not the field. Michael has run thousands of trials, and we now have a long list of chemicals that proved *not* to be the mystery substance. The answer is not far away, and our understanding of the system has improved considerably. As I explained above, we now know that the (small) eggs of invasion-front toads are more sensitive to Bufonite than the (large) eggs of toads from range-core areas in Queensland. Tadpoles from Western Australia also produce more of the chemical. Clearly, the chemical language of tadpoles has evolved just as quickly as the toads' size, shape, behavior, and so forth. Bufonite may not be as useful for toad control in Brisbane as it is in Kununurra—but there's a twist in that tale as well, which I'll explain later in the chapter when I discuss "genetic back-burning." A chemical that kills the eggs of superdispersive invasion-front toads—but doesn't affect the eggs of sit-at-home Queenslanders—may let us manipulate the course of an invasion.

Suppose everything works out exactly as we have hoped. The toad-control methods work, and they work well. Cane Toad populations crash. It's unrealistic to expect toads to be exterminated, but perhaps their numbers will fall to very low levels—say, 10 percent of current numbers. Can we high-five each other and proudly claim "mission accomplished"? No. That achievement, spectacular as it would be, might not reduce the ecological impact of Cane Toads in Australia.

Here's the problem. The reason the Cane Toads are a catastrophe in Australia is that they fatally poison native predators (such as goannas and quolls) that try to eat them. So, predators would be safe if toads were completely eliminated. But reducing toad numbers may not help. If toads become so rare that most goannas never meet a toad, that's well and good. But if we eliminate only 90 percent of the toads, we haven't guaranteed a toad-free existence for those predators. From a goanna's point of view, encountering a single Cane Toad is as deadly as encountering a thousand of them.

And it gets worse. Toads are easy to find. Unlike native frogs, Cane Toads sit out in the open rather than hiding away. The Cane Toad advertises its presence, saying, "I'm poisonous, stay away from me." That's a great tactic in the Amazon rainforest, where evolution has educated the predators about toad toxicity. But in the Australian savanna, especially near the invasion front, the goanna sees a toad as lunch.

So, we have the deadly combination of a predator that is good at finding prey and a poisonous prey item that doesn't try to hide. Even a single toad in the swamp is likely to come to the local goanna's attention. Does that mean that we should forget about toad control? No. Cane Toads don't belong in Australia. But we have to be realistic. We will never eradicate Cane Toads; all we can do is reduce their numbers.

Fortunately, there *is* another way to save native wildlife. Even if we can't eliminate toads, we can toad-proof the natives. We can change the outcome of encounters between toads and predators. Achieving that goal was Team Bufo's first victory in toad control. Indeed, it was the first victory of science in the war against the Cane Toad.

The breakthrough came in 2006, when we exposed live predators to toads in laboratory trials. To my astonishment, many predators that ate toads didn't die; they just became sick, and then they stopped eating toads. The implications of that result should have been obvious, but it took a year for the light to dawn on me. We can exploit that learning ability, by teaching imperiled native species not to eat toads. A toad-smart predator can coexist with Cane Toads instead of dying.

The species we need to teach are the *large* predators. They attack adult toads, a recipe for disaster. A mouthful of supersized *Bufo* provides a fatal dose of toxin rather than an educational experience. The big predators are not stupid. But they never get a chance to learn because all of the toads at the fast-moving invasion front are very large (and, thus, very poisonous).

Why do large toads dominate the invasion vanguard? In evolutionary jargon, reproduction trades off with dispersal. Energy devoted to reproduction leaves less energy for dispersing—you can do one or the other, but not both. Breeding toads are left behind. Added to that is a more practical consideration. In any footrace, a female carrying a clutch of eggs (about 30 percent of her body weight) will lag behind her nonpregnant counterparts. As a result, reproduction is rare at the invasion front. And young toads don't have the speed or endurance to keep up with larger ones, which means that they get left behind as well. The end result is that the invasion front is dominated by large adult Cane Toads—so large that a predator dies after taking a single bite. Quolls, goannas, and snakes are doomed, because the first Cane Toad they meet is big enough to kill them.

The evolutionary acceleration of the Cane Toad invasion through tropical Australia has created a nightmare scenario for vulnerable native species. If that acceleration hadn't happened, Cane Toads would reproduce as soon as they arrived in a new area, and young toads would arrive almost as soon as older ones. As a result, a predator's first encounter might be with a baby toad, not a large adult. That might teach it to avoid toads—so that when a fatally large Cane Toad finally looms into view, the predator leaves it alone.

That scenario suggests a simple way to buffer the toad's ecological impact: *Release small Cane Toads at the invasion front, to give predators an opportunity to learn rather than die.* The idea popped into my head in 2008, while I was preparing a talk to give at the Australian Academy of Science. To my delight, I had been awarded the academy's top biology award—the MacFarlane Burnet Medal—and invited to give a public talk at the academy's annual meeting. This was seriously scary. The academy is the peak scientific group in Australia and occupies an elegant building that (to make it even more frightening) is named the Shine Dome, in honor of my older brother, John. He's an eminent molecular biologist. No pressure on the younger brother. Not much to live up to. Was I prepared to stand up in front of the country's top scientists and suggest that the best way to combat Cane Toad impact was to release more toads?

The idea is simple. If we release small teacher toads just in advance of the main toad front, the predators have an opportunity to learn: Eat a small toad, feel ill, and don't eat any more toads. That might save their lives when the site is overrun by the amphibian invasion a few weeks later. Spreading teacher toads won't do any harm, because the site will be overrun by the Toad Army within a few weeks anyway.

The scientists thought it was a great idea, but it attracted howls of outrage from the toad-busters. The newspaper in Darwin quoted Graeme Sawyer (the head of Frogwatch NT and recently elected lord mayor) as saying, "What planet does this guy come off?" Taking a more personal approach, one KTB member took me aside during a trip I made to Kununurra to meet with the community group. He threatened dire retribution if he caught me releasing Cane Toads in advance of the front. "Lunacy!" he shouted. The idea of teacher toads flew in the face of the toad-busters' claim that they could stop the toads themselves.

The political situation was deeply frustrating. Almost eighty years after Cane Toads were released in Australia, we finally had a simple, practical method that might reduce their ecological impact—but the toad-busters hated the idea. Even if I could convince management agencies to try the approach, it would create a media battle between scientists and community groups. It would be a Pyrrhic victory at best, alienating people whose cooperation would be indispensible for implementing the plan. I didn't want to force people into opposing camps, so I abandoned the "teacher toad" idea for a while. I expected opposition to fade as the futility of hand collecting toads became clear. But, unfortunately, the leaders of those community groups proved surprisingly good at maintaining the rage. It was years before teacher toads were politically acceptable. But one of my postdoc scholars devised a less controversial variation on the theme, so we set to work on it first.

Jonno Webb, who worked with me for many years on snakes, was part of Team Bufo from its earliest days. His new idea about toad control came when he was reading the children's classic "Little Red Riding Hood" to his children at bedtime. Fortuitously, the version that Karlina and Jeremy were hearing included an unusual twist: Little Red Riding Hood hits the Big Bad Wolf over the head with a frying pan, and the grandmother (whom the wolf has eaten) crawls out of his tummy unharmed. The grandmother then carefully cuts the wolf's stomach open and inserts a bag of pungent onions inside. When the wolf recovers, feeling terribly ill, he cries, "I'll never eat grandmothers again."

I'm sure that kids enjoy the thought of the grandmother standing there blinking, covered in wolf vomit. I rather like it myself. But the story had another benefit: it gave Jonno the idea of exaggerating the taste-aversion learning response by adding a nausea-inducing chemical. If a toad makes you a little bit sick, you may forget. But if the same toad makes you desperately ill, you will remember. And, of course, many chemicals that make you sick won't kill you—unlike the toad's poison. So slipping some vomit-inducing chemical into a small toad is safer than giving a large toad to a predator.

To test the idea, we needed access to vulnerable predators—so we could train some but not others, then see what happened after we released them. Fortunately, the Territory Wildlife Park, outside Darwin, was breeding Northern Quolls in captivity. These beautiful creatures—Planigales scaled up to the size of a cat, with a bushy tail and lots of spots—were endangered. Cane Toads were part of the problem, with quoll populations crashing as soon as toads arrived. The Wildlife Park was breeding quolls and releasing the offspring to bolster wild populations but had no idea what happened to the young quolls after release. In 2008 they agreed to let our honors student, Stephanie O'Donnell, educate their captive quolls about Cane Toads. The training would occur in captivity, so the controversial notion of releasing teacher toads into the wild wasn't an issue.

As soon as the baby quolls were ready to forage on their own, Steph offered some of them a small dead Cane Toad (with most of the poison squeezed out of its shoulder glands). She had put a dose of thiabendazole in it, a chemical that causes nausea and vomiting (farmers use it to rid cattle of intestinal parasites). Easier to use than the bag of onions that Miss L. R. R. Hood's grandmother employed, but the idea is the same. The little quolls were quick learners and soon stopped attacking toads. That was encouraging. It meant that quolls were capable of aversion learning, like the smaller marsupials that we had tested earlier. And, even better, when Steph released them into the wild (equipped with radio collars), the educated quolls survived. The ones that she hadn't trained all died as soon as they encountered their first toad.

That difference in survival rates shows that education really does pay off. Steph's results were a shot in the arm for us: finally, we had a way to save an endangered species from the toad invasion. The success of the project struck me forcefully three years later, when I was out collecting Cane Toads one night with Greg and Crystal near a lighted building close to Fogg Dam. The lights attract bugs, and the bugs attract toads. As I was watching Cane Toads gobbling up insects, an adult Northern Quoll dashed out from behind the building, pushed a toad out of the way, and seized a large moth. She was one of our trained quolls: the wild ones had disappeared as soon as toads arrived. It was dramatic evidence that quolls can survive in a toad-infested landscape,

ETHICAL REGULATION OF ANIMAL RESEARCH

The ethical treatment of research animals is a minefield. Socially, it generates a seething mass of good intentions, intuition-based judgments, and hidden agendas. The bureaucracy involved—at institutional, state, and national levels—is nothing short of extraordinary. It would be funny if the underlying issue weren't so important. Like most ecologists, I chose my career because I love animals. As a result, ethical treatment of my study species—including "unpopular" animals like Cane Toads—sits front and center in my values. I'm a benevolent supervisor of my students, but woe betide anyone whose thoughtlessness causes their study animals to suffer. One student made that mistake, and I threw her out of my group.

Most of my colleagues have the same devotion to ethical treatment of animals that I do. But society in general has very jaundiced views of "vivisectionists," and fundamentalist groups push extremist antiscience agendas. In the eyes of many animal liberationists, scientists are malevolent psychopaths who enjoy an animal's pain.

As animal-welfare groups sprang up in the 1980s, they drew attention to occasional cases of horrendous cruelty perpetrated by scientists. Most of those scandals involved research by physiologists, not ecologists, but that distinction was ignored. A tidal wave of antiscience rhetoric spawned a new industry: the ethical regulation of animal research. Farmers were ignored; scientists were the target. At a community market held near my home in Sydney every Sunday, an elderly Chinese man erects dog-eared posters of a terrified monkey staring wildly into space, strapped into a fearsome contraption that holds electrodes implanted in its brain. The gentleman's aim is to collect signatures on a petition to ban animal experimentation. I don't doubt his sincerity, but those pictures are fifty years old, an outdated caricature of what happens today. And although ecologists aren't in his firing line, regulations to curb unethical procedures have had a devastating impact on ecological research—and, thus, have decreased our ability to conserve native animals. That well-intentioned gentleman is part of an antiscience army.

In response to community outrage over barbaric treatment of research animals, authorities created a new bureaucracy. It was done with the best of intentions—to improve animal welfare—but the system soon became unwieldy. Bureaucratic "solutions" to complicated problems (like workplace bullying, racism, sexism, and safety) are important, but if we're not careful we end up with so many rules that nobody can do their jobs anymore.

As an example of the idiocy, what do I need to do if I want to look at a Cane Toad that is sitting beside the road as I drive by? Let's say that I would like to look through the car window and glance at the toad to record its size class and sex, and then drive on without stopping. Other cars on the road are swerving to run that toad over, and community groups are killing thousands of its fellow toads (without needing ethics approvals). But because I'm a scientist, things are tougher for me. A toad might experience stress if I look at it, so I need permission from the university's ethics committee. The application form is ten pages long, will take months to process, and may well be rejected.

My ethics application is far more likely to fail if I want to pick the toad up or attach a radio transmitter to it. To justify my plans, I have to explain the value of the work for conservation. OK, fair enough. But who evaluates my justification? Because biologists are viewed as untrustworthy, ethics committees include many nonscientists who struggle to understand the applications they are asked to assess. The people on these committees are genuine, helpful souls—but the system is diabolical.

The most difficult ethical situation I faced with my Cane Toad research was the idea of putting a live predator together with a live toad. Unfortunately, there is no other way to find out if a predator species is capable of aversion learning. Many people would say that such an experiment is inhumane to the predator (it might die if it eats the toad) or the toad (it might be killed). People object to this kind of experiment even if the predators are collected from areas about to be invaded by toads—and so will be in this situation for real, very soon—and even if they are collected as part of biological surveys (and thus are destined to be killed and preserved as soon as the trials are finished). It just feels wrong, ethically, to put a predator and prey together in the same cage. I understand that feeling. I had sleepless nights when I realized that we needed to run this kind of study.

I was faced with a bureaucratic nightmare as well as an ethical dilemma. Did the conservation benefits of running those predator-prey trials outweigh the suffering of the animals involved? Even if I could convince myself, could I also convince the institutional ethics committee? In the end, I obtained the committee's permission and ran the trials. They suggested a new (and spectacularly successful) way to buffer the toad's impact—by inducing taste aversion. I regret the suffering caused by those trials, but the end result of the work was to spare millions of native animals from Death By Toad. Does that make the work justifiable? I'm not a philosopher. Others can decide.

simply by keeping toads off the menu. Ten seconds of watching that quoll feed on moths, surrounded by toads, was a reward for years of research. That evening, we toasted Little Red Riding Hood at dinner.

But would the method actually save enough quolls or other predators to make a real difference? Or was it just a short-term fix—as soon as those trained animals died, would it be back to square one? Would the young predators in the next generation, not knowing that toads are deadly, try to eat them? If so, what would be the point? Fortunately, the scenario is not that bleak, for two reasons.

The first is that in some species, offspring can learn from their parents. That won't happen with reptiles, frogs, or fishes, because their parents wander away long before the eggs hatch. But quolls are social, and the youngsters hang around with their mother for weeks before they strike out to make a living on their own. So there's plenty of opportunity for a female quoll to teach her offspring that toads are a poor culinary choice.

But there's another, and more general, reason why training a single generation of predators gives a long-term benefit, and it applies to the asocial reptiles as well as the family-living marsupials. The predators at the toad front meet only large toads, so they don't get a chance to learn (unless we help them). But farther back, where toads have been present for several years, they are breeding—so there are small toads as well as larger ones. The pond margins teem with small toads—ideal "teachers" for the offspring of predators. Goannas live long enough to see the invasion-front Cane Toads move on and be replaced by more normal toads that begin to breed. And when that happens, the offspring of those aversion-trained predators hatch into a world containing small as well as large toads. It's now an educational opportunity rather than a death sentence—a school, not a war zone. The young predators can learn toad-avoidance on their own, without any help from us or from their parents.

To look for this kind of long-term effect, we needed a follow-up field study: train Northern Quolls and release them, trap in the same area for the next few years, and test DNA samples from young quolls to see if they were the offspring of the quolls that we released. During her Ph.D. research under Jonno's supervision, Teigen Cremona showed that the second generation does indeed survive. There are toads everywhere, but the young quolls ignore them. It looks encouraging, but two complications have arisen. First, Teigen also caught wild quolls at her Kakadu field site; the rumors of their post-toad extinction had been exaggerated. And, frustratingly, half-wild dogs killed off many of her study animals. Northern Quolls are still in trouble, even if we teach them to stay away from toads.

Can aversion learning save other predators as well? Reptiles are more difficult to train than mammals (I hate to admit it, but snakes are particularly stupid). Still, we've had some successes. Freshwater Crocodiles are top of the class. Young crocs avoid toads after a single noxious meal. That's not too surprising, because crocodiles (despite their fearsome appearance) are more closely related to birds than to snakes or lizards. The crocs have inherited a bird-like, not snake-like, IQ.

We can't give up on "real" reptiles either, because the stakes are too high. In 2011, my student Sam Price-Rees took on the challenge of educating Bluetongue Lizards to avoid toads. The toad invasion front had rolled through Fogg Dam five years earlier, obliterating the lizard's populations, so Sam captured blueys on the outskirts of Kununurra in Western Australia, brought them back to the lab, and offered them sausages that she had constructed from toad flesh. To make the meal more memorable, Sam added a dose of a powerful emetic. Sure enough, the blueys devoured their sausages, became ill, and lost all interest in toad-smelling sausages as a result. Sam then released them back in Kununurra (together with other blueys, captured and handled in the same way but never exposed to the Sausage From Hell) and monitored their survival by radio tracking. Just as for Steph's quolls three years earlier, it worked. Education ensured survival. And, critically, Sam had taken the method one more step toward realistic application in the wild. Instead of using captive-raised babies (like Steph's quolls), Sam showed that conditioned taste aversion (see chapter 6) can save wild-caught predators.

But that still leaves the toad's most high-profile victims. Most lizard-lovers believe that goannas (varanid lizards) are the world's smartest reptiles, based on their erect stance and steady gaze. If so, goannas should be good students and quickly learn not to eat toads. At the same site (Kununurra) and same time (2011) that Sam was studying her blueys, Jai Thomas caught goannas, fed them nausea-inducing sausages, then released and radio-tracked them. But, unlike with the quolls and blueys, training didn't increase goanna survival. The fundamental problem is that these big, bold reptiles are primed to attack anything that moves. Like a Jack Russell terrier, a goanna can't resist a small hopping object. I can sympathize; I have the same reaction to ice cream. It would take a lot of conditioned taste aversion to make me ignore a tub of extra-creamy vanilla.

We know that goannas can avoid eating toads in the field, however, because that's what they do in areas of Queensland where toads have been present for decades. How much of that resistance is driven by genetic changes (adaptation) and how much by learning? We don't know, because this isn't easy science to do. We would need to obtain newly hatched goannas and record their behavior when they first encounter a Cane Toad.

Clearly, sausages didn't create an effective learning experience for a goanna. To switch off that instant-attack response, we needed to educate goannas with an object that looked, moved, and smelled like a real toad. Our best option would be to use teacher toads—the idea I had suggested years before but had to abandon because of community-group resistance. The political landscape had changed enough that I thought a field trial was feasible, especially if we didn't talk about it too much in the media. But even so, it would be a tightrope. We needed toads big enough to make goannas ill, but not big enough to kill them. And we had to conduct those trials in the field, because goannas don't take kindly to captivity. Nobody learns effectively if they are focused on escaping.

I wasn't sure that such a study was possible, but Georgia Ward-Fear took on the challenge for her Ph.D. work in 2013. By then, though, the toad invasion front was moving through almost-inaccessible country in the rocky landscape of the Kimberley. We had to identify a study site that Cane Toads had not yet invaded, but one that they would reach within a year or two at most. After a lot of guesswork, Georgia decided to base her study at Oombulgurri (once known as "Forrest River"), an abandoned Aboriginal settlement west of Wyndham.

The Oombulgurri floodplain is genuinely remote. No roads, so everything (people, equipment, supplies) has to be brought in by helicopter or boat. But it's magnificent country: more arid than the land around Fogg Dam, with spinifex grasses on the higher, drier edges of the floodplain. Despite its grandeur, Oombulgurri has a tragic place in the history of race relations in early Australia. In 1926, an indigenous man (Lumbia) killed a white pastoralist (Fred Hay) who had raped Lumbia's child-wife Anulgoo. A law-enforcement party from the white community exacted savage retaliation. The details are contested, but a royal commission in 1927 concluded that eleven people had been killed in the "Forrest River massacre." Aboriginal accounts record up to a hundred victims. And the slaughter didn't stop there. Tribal justice blamed Anulgoo for initiating the massacre, so she was speared to death.

Oombulgurri was a thriving remote indigenous community from the 1970s onward, with hundreds of residents. But numbers dwindled and social problems increased. The state government closed the settlement in 2011, a few

"Conditioned taste aversion" training on Oombulgurri floodplain in the remote Kimberley, to protect predators from Cane Toads. My student Georgia Ward-Fear and Balanggarra rangers Herbert and Wesley Alberts with a recently captured Yellow-spotted Goanna that will be trained to avoid Cane Toads. Photo by David Pearson.

years before Georgia commenced her studies there. Conveniently, they left some houses and infrastructure, despite the remoteness of the location. Local indigenous people still feel a strong link to the floodplain, so Georgia worked closely with the Balanggarra community as well as the Western Australian Department of Parks and Wildlife.

I was worried that the project was hopelessly ambitious. All the complexities and expenses of a remote site, plus the political sensitivities and the need for cross-cultural collaboration. But no one has ever accused Georgia of being a wimp. Working closely with the local indigenous rangers, Georgia caught and radio-tracked more than a hundred Yellow-spotted Goannas. And when the toad invasion came storming through in 2014, Georgia and the Balanggarra rangers embarked on a vigorous education campaign. They drove across the floodplain on quad bikes from dawn to dusk, radio-locating lizards. When they found them, they crept up quietly and dangled a Cane Toad from a fishing rod, tempting each goanna into tasting (and being sickened by) a small example of the alien toads that were about to pour into the floodplain.

It worked. By the end of the eighteen-month study, only one untrained lizard had survived longer than 110 days, compared to more than half of the trained lizards. Goannas thrived even in areas that were swarming with Cane Toads. Georgia's work had taken the conditioned-taste-aversion approach a critical step further than either Steph or Sam. She trained free-ranging lizards by exposing them to live Cane Toads—not captive animals, not sausages. It wasn't quite the same as releasing small teacher toads on the floodplain in advance of the Cane Toad invasion front—the idea I had originally suggested—but it was close enough. Public opinion about the idea began to change. No longer was it an idiotic fantasy dreamed up by an out-of-touch boffin from a faraway university. Demonstrably, it worked. It was an effective way to save native animals. And the political landscape had changed as well; the local indigenous community had been an integral part of Georgia's study, they had seen it all unfold, and they recognized a spectacular success when they saw it. And, increasingly, the opinions of indigenous leaders were being treated with great respect.

By 2016, when Georgia's paper on goanna education appeared (with the Balanggarra rangers as coauthors), the tide of public opinion had swung firmly toward teacher toads. Even Graeme Sawyer, who opposed the idea initially, was converted to the revolution. Even better, the Western Australian government department in charge of wildlife management—which had worked closely with us during Georgia's project—was convinced as well. In late 2016, the Department of Parks and Wildlife committed to collaborating with us to roll out teacher toads on a landscape scale.

But I was still worried about negative opinions from knuckle-draggers. The idea of "spreading more Cane Toads to save the wildlife from Cane Toads" sounds preposterous, and I feared that many people would react against it without trying to think it through. We might still hit a media firestorm of opposition. But, in an enormous stroke of good fortune (at many levels), fate handed me an opportunity to hammer the airwaves with teacher-toad propaganda.

In an almost-unimaginable coincidence, I was awarded two science prizes in one week in October 2016. And they were the two biggest in the country—New South Wales Scientist of the Year and, even bigger, the top national award, the Prime Minister's Prize for Science. All of the previous winners of the national prize had been scientists in high-profile fields, doing sophisticated research—inventing Wi-Fi, curing cancer, cloning genes, and so forth. The prize had never before gone to an evolutionary biologist or conservationist, let alone a field-worker. I put on a suit and duly attended the New South Wales prize ceremony at Government House in Sydney, then took a flight to

The core members of Team Bufo, far from their natural habitat: Ben, me, Michael, and Greg at Parliament House before the formal ceremony where I was awarded the Prime Minister's Prize for Science in October 2016. The publicity attracted by the award gave me an opportunity to explain the idea of "teacher toads" to a broad audience. Photo by Terri Shine.

Canberra a few days later, to wear my tuxedo while I received the top science prize from the prime minister at Parliament House. The award was a stunning recognition of Team Bufo's achievements, and, fittingly, Greg, Michael, and Ben were all there at the ceremony, in their rented tuxedos. The prize attracted a lot of public attention, giving me an opportunity to spread the idea of teacher toads—and, especially, Georgia's demonstration of its success—far and wide.

In dozens of interviews for radio, TV, and newspapers, I saturation-bombed the media with explanations of the idea and the evidence that it works. Nobody expressed any opposition, at least in public view. With the implicit endorsement of my research by the Australian government, as well as the state governments of New South Wales and Western Australia, the idea of teacher toads had finally achieved respectability. Some of the media reports implied that saving toad-vulnerable predators via conditioned taste aversion was the reason I had been awarded the coveted prize. I didn't correct that interpretation. I was elated. Our research had made a difference. As the Cane Toad invasion rolls inexorably through the pristine wilderness of the

Kimberley over the next few years, thousands of native animals will be spared the agony of Death By Toad.

Like all good science, Georgia's project also shed light on many other issues, not just teacher-toad training. For example, Georgia discovered that goannas were far more abundant on the floodplain than we had guessed; that they grow and mature faster than we had expected; and that, despite their massive size, they are often consumed by pythons. But perhaps the most fascinating result from Georgia's work was a dramatic—and unintended—confirmation of the value of cross-cultural engagement in conservation biology.

When Georgia told me that her individual lizards had consistent personalities—some bold, some shy—I wasn't surprised. Almost all animals exhibit this kind of variation (just think about dogs). And "personality" is a lot easier to notice in a 7-kilogram (15-pound) lizard than in a tiny skink. If your quarry is a giant goanna with formidable weaponry, you really notice if it struggles frantically when you capture it, rather than remaining passive. Or, when you release it, if it attacks you rather than running away. But Georgia took that result much further, showing that a goanna's personality determines its ecology. If we know the personality type of an individual lizard, we can predict the size of its home range, how much time it will spend in different types of habitats, and so forth. For example, shy goannas have small home ranges, live in areas where they are well hidden, and stay away from the habitats where hungry pythons roam.

But the story gets even better. The goanna-catchers worked in teams—one "whitefella" (western scientist) and one Aboriginal ranger—and Georgia recorded who found each lizard. Overall sighting rates were similar between western and indigenous people, but they found different kinds of lizards. Aboriginal hunters are famous for their extraordinary ability to see animals from a distance, and so it proved: when first spotted, "their" goannas were often in thick cover, and far from the quad bike. The western scientists found lizards at closer range and in more open habitats. Sure enough, that translated into "personality"; the "whitefella" lizards were bolder and struggled more when caught, whereas the "Aboriginal" lizards were shyer. And when Cane Toads arrived in the area, the wave of death hit harder on those "whitefella" animals. They were bold enough to try a new menu option, and it killed them. Training was far more effective with the lizards found by the Balanggarra than with the lizards found by western scientists. Already shy, it didn't take much training to convince the indigenous-caught lizards that Cane Toads were best avoided. The outcome of that research program—

Whether a Yellow-spotted Goanna survives the Cane Toad invasion (top) or is fatally poisoned (bottom) depends on both training and the lizard's "personality." Photos by Georgia Ward-Fear and Greg Clarke, respectively.

the demonstration that teacher toads can buffer the impact of the amphibian invasion—was a testament to the collaboration between races. If Georgia had done it the same way as everyone else, relying on university-trained researchers, she would have missed the main story.

Eliminating the threat from Cane Toads won't be enough, on its own, to save Australia's wildlife. But it is an important step, and I sleep better at night knowing that our research has contributed to the survival of tropical predators. Northern Quolls are unbearably cute whirling dervishes, Bluetongue Lizards have been one of my favorite lizards since I was a small child, and Yellow-spotted Goannas are the kings of the floodplain. If ever a fauna was worth saving, it is the apex predators of the Kimberley.

The greatest triumph in invasive-species control would be to eliminate the invader entirely. For Cane Toads in Australia, that will never happen. Instead, I've focused above on Team Bufo's two main achievements—a way to stop toads from breeding, and a way to enable endangered predators to coexist with toads rather than die. But could we do more? Another possibility remains in the realm of speculation: we might be able to prevent the Cane Toad invasion from spreading into a large area they would otherwise reach.

As Cane Toads have spread westward since 1935, they have moved from the equitable climates of coastal Queensland to the seasonally arid landscapes of northwestern Australia. The world around them has changed from green to red-brown. It's no surprise that the initial Cane Toads thrived in coastal Queensland; rainfall and shelter were plentiful, and temperatures were mild. But those benevolent conditions were soon left behind, and the Cane Toad's march took it into drier, hotter regions, where drought holds sway for most of the year.

Toads have overcome those challenges mostly by escaping the hard times rather than adapting to them. In the seasonally dry tropics, Cane Toads move long distances only when monsoonal rains drench the land. The rest of the year, they take refuge in waterside burrows or beside the houses and farm dams that dot the continent, providing moist, cool conditions that enable Cane Toads to cling to life. Outside such a refuge, a toad would die within minutes. As the toads continue to push westward, they are encountering an ever-more-harsh environment. Natural water bodies are virtually nonexistent, houses are rare, and farm dams become the only oases. What happens if we take away that subsidy?

Removing access to farm dams isn't as difficult as it sounds, because the Cane Toad invasion is heading toward a natural choke point of hyperarid land. Nothing will stop the toads moving through the Kimberley to reach the west coast of Australia—an extraordinary diaspora of around 4,000 kilometers (2,500 miles) in just over eighty years. But then, to continue their spread, the toads must reach the Pilbara region that lies to the south. It's warm enough and wet enough for toads to thrive, and it's huge—about 3,000 square kilometers (1,800 square miles). Rocky hills catch the water, providing year-round refuges for amphibians. But between the Kimberley and the Pilbara lies 700 kilometers (400 miles) of desolate, flat, dry land. The Great Sandy Desert pushes to within 80 kilometers (50 miles) of the coast, leaving only a thin corridor of land moist enough to allow a toad to survive. For most of the year, the only water supplies are farm dams and stock watering points. And even these are few and far between—the land is poor cattle country, where ranchers struggle to make a profit.

In 2013, I collaborated with some of my students and former students, and a few other colleagues, to propose a bold plan. If we could render those artificial watering points inaccessible to Cane Toads (by moving watering troughs up off the ground and eliminating leaky pipes), we might be able to prevent further expansion of the toad front. The amphibians will begin to move south along the coastal corridor during every wet season, but the dry stretch is too wide to cross in a single season. They will try, but they will perish. If the arid-land barrier is effective, the Pilbara will be spared the amphibian onslaught.

Is it feasible? We still don't know. Reid Tingley and Ben Phillips drove along the coastal corridor in 2013, checked out the habitat, and asked local residents for their opinions. Most thought it was worth a try. So Reid and Ben organized a meeting of scientists and land managers in Broome—the town at the north end of the dryland corridor—to identify potential obstacles to the plan. To my surprise, the mood in the meeting gradually shifted from opposition to grudging acceptance that it was worth a try. It would be fairly cheap—less than $100,000 a year—to repair and maintain the water sources so that Cane Toads were excluded. And the time is right: many cattle properties are closing down, and the land is being returned to Aboriginal people to manage. Several of those groups plan to cease farming and let the land return to its natural state. They would be happy to turn off the water.

It's a brave plan, and it may fail. Tropical cyclones deposit huge amounts of rain during the wet season, leaving ponds in place for months afterward.

That chain of watering points might allow Cane Toads to cross to the Pilbara. And, inevitably, toads that have stowed away in trucks loaded with mining equipment will breach the barrier from time to time. The plan would work only with continued vigilance and a willingness to deploy local control measures (like light-traps, hand collecting, and tadpole traps) as soon as an incursion occurs. But if it worked, the biodiversity benefits would be massive.

How we could make the dryland barrier more effective? Early in 2014, I had an idea that I thought was terribly clever, but when I asked Ben for his opinion, I discovered that he had had the same idea. And I received a letter suggesting the same plan a few months later, from an unusual source: a young man who was being held prisoner in a Western Australian jail. For all of us, the idea emerged when we were contemplating evolutionary theory—in particular, the concept of "spatial sorting" (see chapter 4).

To recap: Genes for rapid dispersal accumulate at an expanding range edge, making the invasion front move faster and faster. As a result, the stay-at-home Cane Toads of Brazil, Hawaii, and Queensland have evolved into the athletic, fast-dispersing toads of Western Australia. It's been an eighty-year footrace across the continent, and only the quickest can stay at the front. That evolutionary shift is a major impediment to controlling a biological invasion. It would be far easier to deal with a slow-spreading wave than an accelerating one.

Because toads at the invasion front are single-mindedly focused on dispersing, they cover huge distances—in some cases, kilometers every night. In a year, an invasion-front toad can travel at least 50 kilometers (30 miles), halfway across the dryland barrier. But if the toads that reached that barrier were Queensland stock, they would average only 10 kilometers (6 miles) per year. Their chances of getting across, even in an unusually wet season, would be zero.

Could we make this happen by releasing Queensland toads in advance of the main invasion front, before it arrived at the north end of the dryland barrier? If we did this, the Queenslanders would interbreed with the Western Australians, diluting those "disperser" genes. I wrote to Ben in January 2014: "We could release hordes of Queensland toads in a place where the expanding invasion front will encounter a habitat barrier. If we can swamp out the spatially-sorted disperser genome, maybe we can stop the species from crossing the barrier in northwestern Australia?" Ben immediately pointed out another benefit of the idea: the invasion-front toads have evolved to be good at dispersal, but they are poor at competing with other Cane Toads. Toad abundance is low at the front, so competition isn't important there. Queensland toads would outcompete the newly arriving Western Australians, creating a genetic barrier of sedentary amphibians. Ben suggested the term "genetic back-burn" to describe the idea. Firefighters often light small fires to scorch an area, thus preventing the main fire front from moving through when it reaches that site. Perhaps we could do the same with Cane Toad genes.

Next, Ben and Reid Tingley did some mathematical calculations to investigate the concept. In their models, genetic back-burning worked—and worked well. We would have to convince people to release Cane Toads in advance of the invasion front—at first sight, a truly stupid idea—but the success of Georgia's teacher-toad work had already demolished much of the kneejerk opposition. We could even begin using stay-at-home toads for our teacher-toad trials; if any of them survived to breeding age, they would begin to dilute the single-minded invasion philosophy dominating the vanguard amphibians. Because of the teacher-toad work, we had strong support from the indigenous community. Genetic back-burning would be a difficult idea to sell, but not impossible.

We've run some preliminary trials at Middle Point to see if Queensland toads do indeed outcompete Western Australians, and so far it looks encouraging. And a surprising result from our "suppression pheromone" trials— that the eggs of Western Australian toads are far more sensitive to the suppression chemical than are the eggs of Queensland toads—fits in beautifully. If toads of both types breed at the northern edge of the dryland corridor, that differential vulnerability to tadpole-produced chemicals will wipe out many of the (small) eggs with invasion-front genes, while not affecting the (large) ones with stay-at-home DNA. Once again, we can exploit the warfare between Cane Toads to help control the invaders.

If translocating Queensland toads smacks too much of playing God, and risks disease transfer, we could still achieve the slowdown by selecting locally sourced stay-at-home toads to breed from. Even at the invasion front, we find some toads with short legs and sedentary habits. If we breed from them, and not their more athletic brothers and sisters, the progeny we release are unlikely to travel far. From our trials at Middle Point, we know that leg length is related to dispersal rate, and that both leg length and dispersal rate are heritable. So, even among the toads from nearby regions, we can find individuals whose offspring will be far less dispersive than average. That might be enough to make the difference.

The convergence between these ideas is exciting. By releasing small Cane Toads just in advance of the fast-moving invasion front, we could achieve several objectives at once. Teach predators not to eat toads, and so maintain viable apex predator populations. Introduce hypercompetitive slowpoke toads, to transform the onrushing invasion into a slower-moving stream unable to penetrate into areas that it would otherwise reach. In habitats that lack vulnerable native frogs, we might even be able to reduce the growth rates of those teacher toads (keeping them nonlethal for longer) by infecting them with lungworms. And that would also fulfill the KTB's dream of spreading the parasites to the invasion front line, eliminating the pathogen-free honeymoon period that would otherwise occur.

The idea of removing water sources across that dryland barrier is still being talked about; no firm decisions have been made. But the other two developments from Team Bufo's research have moved from the lab bench to the field. Our pheromone-baited traps are hauling in Cane Toad tadpoles across the entire range of the invasive species in Australia. More importantly for the vulnerable native wildlife, teacher toads are happening. Management authorities in the Kimberley—a coalition of government agencies, indigenous groups, and private landowners—have embraced the concept. The idea of spreading more Cane Toads to save the wildlife from Cane Toads, initially seen as idiocy, has achieved respectability.

And so we've come full circle. Eighty years ago, sugarcane scientist Reg Mungomery walked along the shores of the Little Mulgrave River near Cairns, releasing those first small toads and thereby creating an environmental cataclysm. And now, today, land managers and indigenous rangers are releasing similarly small toads on the other side of the continent, to help lessen the impact of Reg's monumental blunder.

As is generally true in life, the best approach isn't simple. If we want to reduce toad abundance, we need to embrace "integrated pest control" by exploiting multiple vulnerabilities. No single method will guarantee success, but a combination of uppercuts from pheromones and left hooks from habitat restoration, together with a sneaky punch from the tadpoles of native frogs, will have the toad reeling on the ropes. He won't throw in the towel—Cane Toads are battlers extraordinaire—but it's hard for the Toad Army to keep fighting when meat ants and water beetles are eradicating the new recruits, and 20 million years of their own program of weapon development has been turned against them.

Let's not kid ourselves. We haven't won the war. Journalists say that there are around 200 million toads in Australia. I don't know where that number comes from—perhaps by staring at a crystal ball and receiving messages from

Reg Mungomery's ghost. But certainly, there are a lot of Cane Toads out there. It's too early to get cocky about our success, or to start shedding any tears for the Cane Toad. I'm not the first scientist who has thought that he could devise a way to control invasive toads. I'm confident, but so were the others. Those previous armies never even made it past the stage of ambitious blueprints for toad destruction. Team Bufo has moved farther. We now have an arsenal of gleaming new artillery. Environmental managers are implementing our methods on a landscape scale: to stop toads from breeding, to buffer their impact on native wildlife, and to prevent their march into the Pilbara. Will it work? Only time will tell.

ELEVEN

What We've Learned

Theories pass. The frog remains.
JEAN ROSTAND, *Inquiétudes d'un biologiste*, 1967

The Cane Toad Wars continue—and skirmishes between toads and people will go on forever. But Team Bufo's research has changed not only the nature of the battle, but also my own perspectives on many issues.

Some of those changes were driven by new information. When I formed Team Bufo in 2005, the orthodox wisdom was that Cane Toads were devastating to all native species and well-nigh invulnerable; our only hope of saving native species was to kill every Cane Toad we could find. We learned otherwise. The effects of Cane Toads have been more complex, variable, and short lived than we had feared. And our hard-won knowledge of toad biology revealed vulnerabilities. Remarkably, the most powerful methods that we developed—teacher toads, attractant and suppression pheromones, genetic back-burning—all involve using Cane Toads to control Cane Toads.

But, on a broader level, Cane Toads taught me to think differently. My decade on the front line changed the way I view the species involved, the ecosystems, and the interplay between scientific research and the broader community. I'll begin with the other protagonists in the Cane Toad Wars. With the benefit of hindsight, how do I see them?

It's like admitting a murder, but these days I view Cane Toads as rather attractive. I find it more and more difficult to understand why Australians hate them. Before I'm expelled from my homeland for un-Australian sentiment, I'd like to point out (in my defense) that my opinion is the mainstream one from an international perspective. The sight of a toad doesn't elicit horror or revulsion from Americans, Asians, Europeans, or Africans; instead, they view toads as quaint little denizens of the local pond.

Part of the problem is that Cane Toads are so damn big. As a result, we get a good look at the toad's face, with its warts and prominent brow ridges.

Native frogs are strange-looking creatures as well, but they are too small for us to realize it. Their heads are almost as large as their bodies, they have stubby little legs, and their unblinking stare can be disconcerting. If we magically expanded a native frog to the size of a Cane Toad, most people who encountered one would run screaming into the bathroom and lock the door.

Another reason why Cane Toads are reviled is the undeniable fact that they are aliens. They don't belong in Australia. Even kindhearted people—the ones who help little old ladies cross the street, and call the wildlife rescue service if they find a baby bird—advocate brutal treatment of toads. The result is a gleeful massacre of animals that never asked to come to Australia. The toad never had any say in the matter. It's *our* fault that Cane Toads are here. Let's try to eradicate them, but there's no point in blaming the toads. One telling example came from Darwin in 2004, when local communities were hysterically fearful of the toad's imminent arrival. The local newspaper ran a huge headline, "They're Here," with a story about a local woman slaughtering Cane Toads that she found in her backyard pond. The woman said, "I just *knew* they were toads, because they were so ugly." The poor little victims were native frogs.

After they have been talking with the toad-busting organizations, journalists who interview me are surprised by my attitude toward Cane Toads. For toad-busters, the Evil Interloper is desecrating all that we hold near and dear. Team Bufo are far less anti-toad—no zealotry, no crusade, no eco-warriors on a divine mission. Once in a while, a journalist will accuse me of *liking* Cane Toads. In fact, there is a spectrum of opinion within Team Bufo. Greg is on one end, with his fondness for the alien amphibians. For the rest of us, the specter of goannas dying in agony isn't far away when we think about the Cane Toad.

One inescapable fact, however, is that the toad has been an amazing research subject. We have made many discoveries that wouldn't have been possible with another species. Even if you hate toads, you have to admit that they are interesting. Studying them has been a revelation. As Greg said to me once, after yet another research breakthrough, "Cane Toads are the gift that just keeps on giving."

I've been accused of falling victim to "Stockholm Syndrome"—the tendency of hostages to eventually take the side of their captors. And a cynic would point out that the toads have been valuable to me. The toad's high public profile allowed me to expand my research program, and that raises a difficult ethical question. My studies on Cane Toads have been personally rewarding, to a degree that rarely happens in ecological research. It's difficult to hate your sugar daddy, even if he has warts. The science has gone so well that Team Bufo has been showered with national and international awards for research excellence. My career has prospered, far beyond my expectations. If that hadn't happened, would I still feel as positive toward toads? I think so. The change in my attitude came from spending time with Cane Toads, and uncovering their mysteries, not from publicity and awards.

Nowadays, as I walk down the wall of Fogg Dam in the evening, it doesn't annoy me to see a big Cane Toad squatting beside a paperbark tree. I understand that its presence doesn't spell doom for the local fauna, although I bitterly regret the scarcity of goannas and Bluetongue Lizards—undoubted victims of the toad. And the loss of the giant King Brown Snakes cuts very deep. For me, they are a spectacular symbol of the tropical wilderness. But although toads were a factor in their demise, the problems for kingies started long before the Cane Toad arrived. Like the quoll, the kingie was unable to resist the waves of habitat degradation that have swept across tropical Australia.

I feel no personal enmity toward that Cane Toad on the dam wall. It didn't mean to kill native predators, and it isn't here of its own free will. It's just another animal that came to Australia, as many others have in the millions of years before. It is simply trying to make a living in a new land. In the process, the toad has changed. Cane Toads at the invasion front no longer look or act like their ancestors that were released on the Queensland coast in 1935. Cane Toads are becoming Australian. Will we one day view the Cane Toads of Oz as a new kind of animal, playing a powerful role in the ecosystems of its adopted home? We can already see glimmerings of that cultural transition in Queensland, where the toad has been present throughout the lifetimes of most people living there.

What is a "native species" anyway? That's not as stupid a question as it sounds. The Platypus certainly qualifies as a native Aussie, because its ancestors were here 100 million years ago. But how about the Brown Tree Snakes that swam across from New Guinea 100,000 years ago? When Brown Tree Snakes arrived on the small Pacific island of Guam in the 1950s, the local birds were wiped out. The first Brown Tree Snakes that reached Australia may have inflicted similar ecological carnage, but nobody was here to see it. Should we call the Brown Tree Snake a native species? Or do we maintain the rage and regard it as a devastating invader?

If the Brown Tree Snake qualifies as a dinkum Aussie because it found its way here without human help, how about more recent arrivals that came with people? The Dingo arrived less than 5,000 years ago. Is it part of the Australian ecosystem, or do we blame it for wiping out Tasmanian Devils and Thylacines ("Tasmanian Tigers") from the mainland? I'd prefer an Australian continent with marsupial predators, but these days the Dingo is part of the solution, not part of the problem. Dingoes control devastating feral predators like foxes and cats.

How about the Black Rat, which arrived in the 1600s? It displaced native rodents but also took over some of their ecological roles, such as pollinating trees. And it's a terrific Cane Toad "control officer" in southern cities. Should we welcome it? Or consider the Water Buffalo, brought over from Indonesia in the 1800s—like the Cane Toad, it has been a boon for some native species (buffalo wallows are great frog ponds) but the kiss of death for others.

OK, you say, enough of this nonsense. Any species that people brought here is an alien. Get rid of them, and restore the native fauna that lived here before humans wiped them out. But what about Lions in Europe, present in Transcaucasia until the tenth century? Should we bring them back? Do you want to see a hungry Lion outside the boulangerie when you buy your morning croissant? It's not easy to go back to the good old days.

Is the aim of conservation to refashion wilderness as it was before people ruined it? That's nonsense in Africa, where humans evolved. But it's not as much of a problem in Australia, where *Homo sapiens* has been present for only about 60,000 years. Should we eradicate the animals that people moved to new places, especially if the aliens are causing ecological damage? If so, prepare yourself for public uproar when you begin culling Koalas. From eighteen individuals introduced to Kangaroo Island in the 1920s, Koala numbers have exploded, and they are killing the island's trees—but there's no sign of concerned citizens forming a "Stop-the-Koala Foundation." The Aussie "native bear" is too cute.

Conservation isn't as simple as it sounds. What are we trying to achieve? Do we want to go back to some arbitrary time when everything was "natural"? If so, when? Before humans reached Australia? If we had the technology to bring back the giant marsupials, should we? Would it be fair to the species that have replaced them? There are no easy answers. For those of us old enough to have experienced a more pristine world, our memories are tinged with a sense of great loss. I remember the early years in Kakadu, with Northern Quolls bouncing against the fly screens and catching bugs as we ate dinner; and the thrills of sneaking up on a lordly goanna or of wrestling a giant King Brown Snake. Cane Toads are part—but only part—of the reasons why I no longer have those opportunities.

Our emotional investment in nature is critical. It motivates us—professional scientists as well as concerned citizens—to hurl ourselves into the battle against environmental degradation and species loss. But that emotional involvement has costs as well as benefits. It can encourage judgmental thinking and artificial distinctions between "good" and "bad" species. It can encourage us to focus all our energies into fighting against a perceived enemy, while ignoring wider problems. The community outrage directed against Cane Toads was partly warranted and partly misplaced.

One of the things we have learned about Cane Toads in Australia, for example, is that their impact occurs primarily at the invasion front. Large predators are slaughtered as soon as they first encounter toads. The consequences of that impact reverberate for decades—the apex predators are gone—but the massacre itself is short lived. Which leads to a very disturbing question: What's the point of trying to eradicate Cane Toads after they have become established in an area? People spend vast amounts of money and effort on doing this, but the evidence for any benefit to the ecosystem is vanishingly small. After the initial slaughter of top predators, Cane Toads don't seem to have much impact. They may well reduce the number of insects, but (because they live in degraded habitats) they don't compete much with other insect-eating species. Does the continued presence of Cane Toads delay the recovery of apex predators? Perhaps. Does it affect people's peace of mind? Absolutely. But how important are those effects when we sit down to identify priorities for spending scarce conservation dollars?

Even at the invasion front, the Cane Toad's effects are complex. Their impact on large predators is sickening; on the other hand, that Armageddon makes life safer for smaller species. Ironically, the invasion of Cane Toads has benefited more native species than it has imperiled. But is this really a good thing for the system as a whole? Surely toads have destabilized the ecosystem by removing apex predators? Goannas once were common, and the ecosystem has been brutally altered. Toads are a pestilence, pure and simple.

Unfortunately, the situation is not so straightforward. During my fieldwork in the 1980s in Kakadu National Park, Yellow-spotted Goannas were rare because of intense predation by Aboriginal hunters. Whenever we traveled around Kakadu with indigenous people, any goanna that we saw became the target of a spontaneous hunt. And if we had dogs with us, the outcome was always the same: a dead goanna for the campfire that night. This relentless hunting reduced goanna numbers. Plausibly, goannas were common 60,000 years ago (before people arrived in Australia), then were decimated by Aboriginal hunting, then increased again as Aboriginal people moved to the cities in the mid-twentieth century, and then decreased again when Cane Toads arrived a few decades later. What part of this cycle is "natural"? What is the "best" goanna abundance to aim for in conservation terms? There are no easy answers. In a dynamic system, the impacts of Cane Toads are just one more disruption.

And that dynamic plays out at every spatial scale and in every time frame. One of the most profound messages from our Cane Toad research is the importance of rapid evolutionary change. You can't understand the ecological impact of toads in Australia if you treat the invaders as static entities. The amphibian immigrants have become more adept both at rapid dispersal and in dealing with novel environments. Those evolved changes in Cane Toads massively affect their impact. For example, without spatial sorting to drive down reproductive rates of Cane Toads in the invasion vanguard, predators in a just-invaded area would encounter juvenile as well as adult Cane Toads and, thus, have an opportunity to learn taste aversion rather than die. Equally, the duration of toad impact is determined by how quickly native predators adapt to the newcomers, through either learning or natural selection. Large predators are virtually extirpated at the invasion front, but their populations recover within decades. Coexistence with Cane Toads, initially impossible, becomes the norm. Ever since 1935, all the players in this continent-wide drama have been changing in response to the pressures that sprang up when Reg Mungomery released juvenile Cane Toads in cane fields along the Queensland coast.

The key issue here is that ecosystems change through space and time, responding in complex ways to new challenges like global warming or species invasions. Sometimes components fail: the new pressure is too great, or too sudden, and a species is lost forever. More often, the challenge is met with resilience. Some species decline, others increase, and the change is not just about numbers: the attributes of organisms change as well. Processes like spatial sorting and natural selection shape new kinds of animals and plants, different from those that came before. Given the opportunity, ecosystems are supple.

But there's no room for complacency. Despite their resilience to change, natural systems are in crisis. Rates of species extinction have skyrocketed worldwide. The equation is simple: the number of people on our planet is increasing rapidly, putting immense pressure on the Earth's ecosystems. More and more people are plundering a finite store of resources. Ecologists can't stop that degradation; all we can do is buffer it, try to conserve species at risk, and convince our own species to come to its senses before it's too late.

The resilience of the Australian wildlife in the face of toad invasion is one of the most heartening results from our work. When I began my research on toad impacts, I saw our native species as victims: vulnerable losers that would be struck down by the mighty Toad Army. Perhaps there's a hint of Aussie cultural cringe in there. For many years, biologists viewed Australia as a refuge for second-class species that couldn't survive the hurly-burly of life in the real world. Australia was a lifeboat for lesser beings. In that view, invaders from More Important Places would overwhelm the incompetent natives. The Australian species would fall in droves, and the northern immigrants would supplant them.

During my professional life, that view has been obliterated. We now understand that Australian animals are just as complex and sophisticated as their Northern Hemisphere counterparts, and that they can give as good as they get. They aren't inferior, they're just different. Australia has exported its fair share of invaders to other countries, including the Green and Golden Bell Frogs that currently enjoy the hospitality of New Zealand, New Caledonia, and Vanuatu. In the same way, earlier views of amphibians and reptiles as "primitive" or "lower" have been swept away. The ectotherms are different from mammals and birds, but they are just as good at doing what they need to do.

If I'd given that paradigm shift more thought, I would have been less surprised when the Australian wildlife proved resilient to the invading Cane Toad. Yes, there was huge mortality of a few species—but even those victims began to recover after a few decades. And many species changed to become less vulnerable, or even to exploit the toad. When I began my research, I assumed that the toad's impact would be uniformly negative. If you had told me that Cane Toad invasion would increase the abundance of most reptiles, I would have doubted your sanity. But that's what's happened, at least at Fogg Dam.

Even in the nitty-gritty questions of which species adapted to toad invasion, and how quickly they did it, many of my preconceived notions turned out to be wrong. It's a good example of just how flawed your (or, at least, my) intuition can be. I have spent a long career looking at animals, thinking about them, and trying to identify what the hell they are doing. I'm an eminent Scientific Guru; my office walls are adorned with awards and trophies. But when it came to predicting how native reptiles would react to the arrival of Cane Toads, I did about as well as a seven-year-old would do on a university exam in quantum physics. And when Ben, Greg, and I first looked into the crystal ball of ecological theory to predict toad impact, we fell flat on our faces.

There are two take-home lessons here. First, it pays to be humble. Anyone can make mistakes. The second message is that "common sense" can be a poor guide. In this regard, I sympathize with the toad-busters. Some of them spend a lot of time in the bush, see a lot of animals, and know a great deal. But they have opinions and impressions, not evidence. Ironically, my initial opinions were similar to those of the toad-busters. I expected the Cane Toad invasion to be a catastrophe. I changed my mind as the data rolled in, but the toad-busters didn't. That's the primary difference between evidence-based and faith-based approaches—do you incorporate new information or ignore it?

What have we learned about how to deal with an environmental catastrophe? One of the first lessons (but hardest to implement) is to keep your eye on the real issues. Far too much energy in the war against Cane Toads was spent in rivalries between the community groups, and that sometimes spilled over into rivalry between community-group leaders and scientists. It's only human to take offense at criticism, but ultimately it's a waste of time to worry about petty conflicts or to blame people for the positions they have taken.

It's tempting to take umbrage at the men who foolishly brought the Cane Toad to Australia, but apportioning blame doesn't help solve the problem. No heads rolled as a result of that awesomely bad decision. Mungomery and his mentor, Arthur Bell, went on to have successful careers with the Bureau of Sugar Experiment Stations. Bell became director of the bureau in 1945 and died at his desk (of a heart attack) in 1958. The University of Queensland still awards the annual A. F. Bell Memorial Medal for agricultural science. Mungomery rose to assistant director in 1964, retired in 1968, and died in 1972. And although he lost the battle, the hero of the piece—Walter Froggatt—has been remembered not for his valiant opposition to the toad's release, but for other parts of his life. Froggatt Crescent in the Sydney suburb of Croydon bears his name, and a scenic lookout at Ball's Head (a reserve on the Sydney shoreline) is named in his honor. The Froggatt Prize for Science at the Presbyterian Ladies' College in Sydney (the school attended by Froggatt's daughters) is also named for this redoubtable pioneer, but he deserves more. Some organization out there should offer an annual Froggatt Award, for scientists who dare to speak their minds.

The futility of defending the toad eventually dawned even on its former supporters. By 1942, Mungomery was losing hope: "grub damage has increased this year in areas where the toad population is high . . . the number of beetles destroyed by the toad is relatively small. . . ." The toads became yesterday's news after James Buzacott (who built the toad's pleasure palace at Maringa in 1935) discovered another way to kill cane beetles: powerful organochlorine insecticides. These took any political heat out of the toad's failure. By the time people realized that the chemicals caused cancer (and, hence, had to be withdrawn from use), the toads had slipped under the political radar. The Great Toad Plan was quietly put into the category of yesterday's ideas, best not mentioned in polite company. If you visit Maringa today, the scientists at the Sugar Experiment Station don't want to talk about the dismal failure of gigantic American amphibians to save the cane crop.

The leaders of toad-busting community groups have also receded from the public spotlight. As the toad invasion roared on unabated, these groups declined in influence. Nonetheless, they are still a force for good. Most toad-busters are well-intentioned citizens with a genuine love for nature. Despite occasional skirmishes with them, I have more in common with the toad-busters than I do with the property developer who bulldozes a native forest to build a shopping center. And the toad-busters' enthusiasm got people up and out into the bush, away from the TV and the bottle of beer. Toad-busting provides a nucleus for social cohesion and has even built bridges between indigenous and European communities. Toads haven't solved any social problems, but they have built a few friendships.

The toad-busting groups kicked some goals in the environmental dimension as well. They brought the impacts of invasive species to national attention. Some of my scientific colleagues resent the millions of government dollars that went to toad-busters, but equal amounts go to sporting and social clubs. In that sense, the money was well spent. It brought people together in regional Australia and encouraged them to get out into the bush and have some exercise. Cane Toads have been a political football in more ways than one. And the media furor that the busters raised about toads—although based on hyperbole—lifted the public profile of our research. It encouraged governments to increase my funding. Trapped within a system that rewarded inflated claims, the toad-busters were victims of the political process, not perpetrators of evil.

The role of community groups is intricately linked to the role of governments. State and federal governments funded the fight against toads, and some government scientists were active players. For example, the Western Australian Department of Parks and Wildlife employs our collaborator David Pearson, and scientists and wildlife managers from the department have played a key and constructive role in understanding and buffering the impact of toads. Team Bufo couldn't have extended our work into Western Australia without the department's support. Local knowledge and local infrastructure were critical. Of course, some of the governmental responses to the toad invasion—in Western Australia as well as elsewhere—were driven by populist pressure rather than scientific information. For example, the department at one stage employed teams of professional toad-killers to find and destroy the invaders. At many levels of government, I saw intelligent people who were clearly aware of the futility of picking up toads, but the expenditure of funds was driven by political priorities. A marginal electorate attracts more cash, no matter who is in power.

That intersection between science and politics is core to the problem of dealing with environmental issues. If politicians are to be reelected, they need to please the people—even if that means squandering money on futile approaches. Tragically, most environmental problems take years to solve—well beyond the next election. It's important for a politician to look as if he or she is doing something about the problem, but it's not important (electorally) for them to solve it. And if they do manage to solve the problem, it's no longer a hot-button issue for the next election, so there's little political gain in claiming that victory. It's yesterday's news. Any investment that doesn't yield an immediate benefit is likely to move down the list of political priorities. I don't see any way out of this mess, short of maintaining the rage. Environmentalists need to keep shouting, as loudly and often as they can, in as many forums as they can, that our custodianship of the natural world warrants a high priority as we plan for the future.

That plea for priority will be even more effective if science and the community are aligned—if the public understands the benefits of rationally defined programs rather than populist sloganeering. But again, the challenge is formidable and getting worse. Scientific research once held a special place in society, but that's no longer true; scientists are just voices in a media scrum and rarely make their opinions heard above the uproar around them. Evidence-based conservation will work only if scientists can form effective alliances with the broader community.

The Cane Toad wars are both encouraging and discouraging from that point of view. Researchers and community groups sometimes worked together, sharing information and pressuring governments to make evidence-based decisions. But the two groups also spent time at cross-purposes and political point-scoring. Worryingly, that latter mode is emerging as the way of the future. Our society is moving toward conflict rather than cooperation among the groups who value the natural world. The most spectacular examples are increasingly bitter battles between advocates of animal welfare and proponents of science-based conservation. Both groups are well intentioned, and both are motivated by a deep regard for the natural world. But they differ profoundly in how they view the issues, and middle ground is difficult to find. Should we eradicate feral rats from an island where endangered seabirds breed? Yes, say the scientists—the seabirds are facing extinction. No, say the animal-welfare activists, a rat has as much right to live as a seabird; humans have no ethical authority to make judgments about the value of one life versus another.

Those arguments reflect diametrically opposed views of the world. On one side, ecologists look at populations and aim to keep them viable. If we need to kill a few individual animals to obtain the information needed to save an entire species, so be it—the end justifies the means. On the other side of the debate, animal-lovers cherish the individual organism and can brook no harm to it—the end can never justify the means. I believe that population-level thinking offers the only hope of slowing the biodiversity trainwreck that is happening all around us, but scientists are losing that war. Popular opinion is swinging more and more behind the advocates of emotion-based rather than evidence-based approaches to conservation.

Over recent decades, the public power of animal-rights activism has grown. Most people live in cities, shielded from the day-to-day realities of life and death in nature. Even as the city-dweller enjoys his lamb chop for dinner, he is horrified by the thought of animals being killed. Caring about populations rather than individual animals seems callous. And our political leaders keep a close eye on the public's attitudes toward these topics, when they calculate the electoral costs and benefits of supporting the researchers versus the bunny-huggers.

Ironically, although emotional responses impede many conservation initiatives, the reverse was true in the Cane Toad Wars. Revulsion of toads by the public pushed toad-related issues to a high priority on the conservation agenda. Other issues that deserved more attention, like the dramatic collapse of small-mammal populations across the Australian tropics, went unnoticed. The general public doesn't see secretive small mammals too often and doesn't know how to identify them even if they do. All rat-like objects are called "rats." So, nothing obvious had changed, unless you went out looking for Phascogales, Bandicoots, and Brushtail Possums and found that they were no longer there. And, to make it even more difficult for scientists to arouse public concern, the reasons for the near extinction of tropical mammals were (and are) unclear. Easier by far to arouse passion against an ugly amphibian that turns up in your suburb and poisons the local wildlife.

In other biological invasions, attempts to curtail the invaders' spread, or to reduce their numbers, have generated vigorous opposition from the general community. In Hawaii, vigilante groups intentionally spread an invasive frog called the "Coquí" around the islands. The Hawaiian legislature imposed massive penalties—up to $200,000 and three years in prison for anyone intentionally transporting Coquís! But it was too late. The frogs were already widespread, rendering eradication impossible. In northern Italy, attempts to cull invasive Gray Squirrels were abandoned after public concern (and, eventually, legal action) over the idea of violence against cute, furry creatures. In Australia, wild horses continue to degrade alpine habitats because well-intentioned horse-lovers cannot countenance the thought of killing a horse—any horse. If the public saw Cane Toads as cute—if they were bright green, for example, instead of brown—the Cane Toad Wars would have played out very differently.

This is the dynamic societal landscape in which decisions about conservation are made: a populace that is increasingly divorced from nature, with little understanding of ecological processes; and a political machine focused on short-term crises rather than long-term threats. To make our leaders respond effectively to any ecological issue, we first must convince the general community that there is a problem. That is a difficult task, with a thousand other "problems," both real and imagined, competing for public attention every day. Next, we need to go further and convince the community that this problem is best viewed from a population perspective—again a difficult challenge, because appeals to focus on the suffering of individual animals will be difficult to counter. And third, we need to suggest a rational way forward, to help solve the problem. Unless we have brought the community with us every step of the way, our chances of success are slight.

The obstacles to a science-based approach to conservation, then, are as much about community involvement as scientific breakthroughs. Finding a way to solve an environmental problem is only the first step. Building a broad coalition within the general community, supportive of a science-based approach, is the next hurdle—and the one at which many scientists have foundered. To construct those bridges, scientists will need to redefine their own roles.

Last but not least, the War of the Toad has led me to a full-blown love affair with my main study site, Fogg Dam. The floodplain of the Adelaide River is not as spectacular as the sandstone escarpments of Kakadu National Park; there are no streams, waterfalls, or Aboriginal rock art. To a newcomer, it looks like clapped-out buffalo country: a few paddocks among the eucalypts, and patches of monsoon forest near the water. The floodplain couldn't be plainer. Occasional small ponds, scattered clumps of reeds, but otherwise a vast, flat expanse, peppered with deep cracks in the soil for most of the year. After the fires come through—and they do, every year—it is a charred moonscape.

But looks aren't everything. The floodplain holds an extraordinary diversity and richness of wildlife. It doesn't give up its secrets easily. It's taken decades for people like Greg to peel back the layers and reveal the stories beneath. But that's part of the charm. We've had to work hard on our relationship with this place. We now understand the Fogg Dam ecosystem with an intimacy that we can claim for very few parts of the planet. And, just as a decades-long relationship between two people brings with it a depth of feeling, so does the opportunity to see this magnificent ecosystem in all its moods.

As in any relationship, there have been bad times as well as good: hypotheses that didn't work out—or, worse, that we failed to test properly—and, with encroachment by humans and their feral pests, a growing sense of an ecosystem under pressure. The arrival of Cane Toads, although it launched me into an exciting new phase of my career, was catastrophic for creatures that I care about. I have forgotten every staff meeting I ever attended in Sydney, but I can remember every King Brown Snake I've seen on the dam wall. There was an excitement to being the first reptile ecologist in tropical Australia. There's a different joy to being an old professor. I'm proud of my role. I haven't done too many of the hard yards out there on the floodplain, bitten by mosquitoes and chased by crocodiles, but I've done enough to have a flavor of it all. And I've helped other people, better field-workers than I could ever be, reveal the functioning of the floodplain ecosystem in all its awesome complexity.

When I'm back in Sydney, stuck in a meeting or listening to a seminar, my mind often wanders to the wall of Fogg Dam: dusk settling in; the night shift of wildlife beginning to stir; the sights, sounds, and smells of the floodplain. We have no idea what we'll see tonight. It's a perennial lottery, but there is always something to store in the memory banks and bring out from time to time to relish—for its grandeur, its humor, the new thought that it sparks, or the hypothesis that it supports or refutes. Or just the feelings that it evokes. Bewitching.

But not "pristine." This is a landscape sculpted by people. Except for the floodplain, the land is cultivated for bananas, mangoes, and melons, or grazed by buffalo or cattle. So I'm not a lovesick youth, swooning over a sweet, virginal girl next door. The object of my affections has been shaped by her previous relationships. It's tempting to see the current state of the landscape as "natural," but I've seen Fogg Dam change year after year. I've seen it transform from open water to tea-tree forest after the Water Buffaloes were culled; and the endless dry-season jungle of Speargrass changed into a woodland mosaic by a shift in fire regimes. I've seen the towns expand, changing from frontier villages to modern suburbs. I've seen the forest cleared for farms, and sealed roads replace the old dirt tracks.

A few years ago, the national parks service dug ponds in the floodplain below the dam wall to attract birds during the dry season. When the bulldozers dug down, they hit well-preserved mangrove roots, only a few thousand years old. Just a moment ago, on a geological time scale, this was the coastline. The sea level has fallen, leaving Fogg Dam high and dry. The ecosystem that I have studied is a recent phenomenon. Every animal there is an invader—a species that moved in as the sea level fell. The rats and Water Pythons are "native" only because they arrived before the Cane Toads. Seen through that prism, it's difficult to view the toad as truly different. Our fond ideals of calm and stable ecosystems are nothing but wishful thinking. Change is the only constant.

Cane Toads are like those irritating people who bought the house next door and opened a tattoo parlor. When they arrived, they caused problems. Big problems. And some of the locals are gone as a result. But, with time, we begin to see the interlopers in less black-and-white terms. We can learn from the toads. For many people, Cane Toads will forever be the evil aliens, responsible for the progressive decay of the ecosystems around us. It's comforting to have someone to blame. But my own resentment has declined. I wish those new neighbors hadn't moved in, and I'm doing all I can to minimize their bad effects. We have to live with them, and we can. The good news is that the ecosystem has handled these new arrivals better than we ever thought it would. Even with Cane Toads on the dam wall, Fogg Dam remains a magical place.

Stepping back a little, what have we learned from the War of the Toads? Are there broader lessons here, about how to deal with a biological invasion? I think there are. In purely pragmatic terms, the toads remind us that quarantine is the most cost-effective way to deal with invaders. Once they are here, they are here to stay. But we don't have to throw our hands in the air and give up. Embarking on a zealous attempt to massacre the invaders is unlikely to pay off, but investment in research—to understand the enemy—may produce novel weapons that can turn the tide in our favor. Beware simple solutions, and look closely at the credentials of people who offer them. We need solutions that work, not words that make us feel good. Avoid knee-jerk responses, and don't assume that the alien is a problem just because it's an alien. Targeting the impacts can be more effective than targeting the alien itself.

Some facets of the Cane Toad Wars revolved around the toad itself, and others were based in Australian rural communities and their attitudes. But much of what happened as the foreign amphibians hurtled through Australia is a microcosm of what happens in bigger environmental issues. That's true not just for the biology of the problem, but also for how we deal with it. We can learn from how scientists, politicians, journalists, and the general public viewed the challenge of the toad, and from how they reacted. Some things went well, and some could have been done better. Resources were wasted in futile efforts, while productive options were ignored. But, in the end, the main message is a positive one. Enough people cared about the ecosystems of tropical Australia. Despite different agendas, we were able to work together to build an understanding of what was going on—and, thus, to help the ecosystem cope with a problem that had once seemed insoluble. At the end of the day, the people of Australia cared enough about conservation.

Environmental challenges will keep coming. We create them, and we need to fix them. If we can learn from the Cane Toad Wars, we can do a better job of dealing with the next crisis. Cane Toads are here to stay—but probably at lower numbers, and with less impact on native fauna, than has been the case over the past eighty years. There will be times and places where toads are abundant, but it will become the exception rather than the rule. We can take some reassurance from the resilience of the Australian wildlife in the face of the toad invasion, but the signs are ominous. If we keep trashing our unique ecosystems, how much longer will they be able to deal with wave after wave of new challenges?

ACKNOWLEDGMENTS

First and foremost, thanks to Team Bufo. Greg Brown combines world-class field biology with an ability to conduct sophisticated immunological analyses—and then fix a malfunctioning toilet. He still enjoys walking along the Fogg Dam wall every night, after already covering (by my calculations) more than 20,000 kilometers (12,000 miles) in the process. Ben Phillips can seamlessly switch from radio-tracking Death Adders to running spatially explicit mathematical models. And if you want to set up a perfectly balanced, exquisitely designed experiment, Michael Crossland is your man. I must also acknowledge Jonno Webb's contributions (and his superhuman skill at catching more fish than I do).

For almost twenty years, the world's best technical officer, Melanie Elphick, has supported me—with a smile that just keeps getting bigger. Cathy Shilton's expertise with pathology has been exceeded only by her ability to create mango daiquiris. Other Team Bufonids have done amazing things also, whether they be postdocs (Ligia Pizzatto, Mattias Hagman, Jason Kolbe, Matt Greenlees, Sylvain Dubey, Adele Haythornthwaite, Tom Lindström, Takashi Haramura, Camila Both, Jayna DeVore, Simon Ducatez, Lee Ann Rollins, Mark Richardson, Chris Friesen), graduate students (John Llewelyn, Christa Beckmann, Ruchira Somaweera, Samantha Price-Rees, Crystal Kelehear, Reid Tingley, Elisa Cabrera-Guzman, Edna Gonzalez-Bernal, Joshua Amiel, Uditha Wijethunga, Georgia Ward-Fear, Dan Natusch, Cam Hudson, Jodie Gruber, Sam McCann, Georgia Kosmala, Greg Clarke), honors students (Haley Bowcock, Travis Child, David Nelson, Stephanie O'Donnell, David Llewellyn, Iris Bleach, Damian Lettoof, Chris Jolly, Felicity Nelson, Damian Holden, Milly Raven, Lachlan Pettit, Renee Silvester, Kat Stuart, Patt Finnerty), or technical assistants (Michelle Franklin, Nilu Somaweera, Chalene Bezzina). And each of the volunteers, parents, siblings, partners, and friends deserves their own paragraph, but there's just no room. Eric Cox, OAM, the buffalo farmer who helped me build a research base in the tropics, deserves an entire page.

Modern science is all about collaboration. There's too much for any one person to know. So take a bow, David Pearson (Western Australian Department of the Environment), Ross Alford and Lin Schwarzkopf (James Cook Uni), Keith Christian and Di Barton (Charles Darwin Uni), Lee Ann Rollins (Deakin Uni), Rob Capon and Angela Salim (Queensland Uni), Dave Newell (Southern Cross Uni), Dave Skelly (Yale Uni), and my hordes of other coauthors. Thomas Madsen's enthusiasm, creativity, and confidence were critical in getting my tropical research up and running. Our access to study areas was due to the kindness of Ken Levey, Grant Hamilton, Les Huth, and Graham Kenyon. The Fogg Dam rangers—Barry Scott, Mark Fogarty, Sam MacKenzie, Adrian MacKenzie, and Dean Lonza—supported the crazy toad people, and our neighbors at Middle Point—Heather, Jerry, Dave, and Lucy—let us trample around their water bodies.

The unsung heroes of ecological research are the funding bodies. Without the Australian Research Council's support, our work wouldn't have happened.

My home base—the University of Sydney—deserves credit for supporting me, as do colleagues like Gordon Grigg and Mats Olsson. Among a host of partner organizations, the Western Australian Department of Parks and Wildlife and the Australian Reptile Park are the first cabs off the rank. The Northern Territory Land Corporation has been a benevolent landlord ever since we took over Middle Point Village.

Eric Engle, Developmental Editor Extraordinaire, turned the tedious and poorly organized first draft of this book into something more intelligible. Harry Greene kindly wrote the foreword.

Most of all, I thank my family: my wife, Terri, and sons, Mac and Ben. If our work ever manages to make a real difference to toad control in Australia, you guys need to watch out for vigilante teams of Cane Toads. You're the reason for any success that I've ever had. If the toads find out, you need to be careful. Cane Toads don't take prisoners.

APPENDIX

Species Common and Scientific Names

African Clawed Frog, *Xenopus laevis*
Agile Wallaby, *Macropus agilis*
American Bullfrog, *Lithobates catesbeianus*
Anaconda, *Eunectes murinus*
Antilopine Wallaroo, *Macropus antilopinus*
Arafura Filesnake, *Acrochordus arafurae*
Asian House Gecko, *Hemidactylus frenatus*
Bandicoot, *Perameles nasuta*
Barramundi, *Lates calcarifer*
Black Kite, *Milvus migrans*
Black Rat, *Rattus rattus*
Black-spined Toad, *Duttaphrynus melanostictus*
Black Whipsnake, *Demansia vestigiata*
Bluetongue Lizard, *Tiliqua scincoides*
Blue-winged Kookaburra, *Dacelo leachii*
Brown Rat, *Rattus norvegicus*
Brown Tree Snake, *Boiga irregularis*
Brushtail Possum, *Trichosurus vulpecula*

Burton's Legless Lizard, *Lialis burtonis*
Cactus Moth, *Cactoblastis cactorum*
Camel, *Camelus dromedarius*
Cane Toad, *Bufo marinus* [or *Rhinella marina*]
Carp, *Cyprinus carpio*
Cat, *Felis catus*
Chestnut Blight Fungus, *Cryphonectria parasitica*
Chimpanzee, *Pan troglodytes*
Chytrid Fungus, *Batrachochytrium dendrobatis*
Coquí, *Eleutherodactylus coqui*
Corella, *Cacatua sanguinea*
Corroboree Frog, *Pseudophryne corroboree*
Crimson Finch, *Neochmia phaeton*
Crimson-spotted Rainbowfish, *Melatotaenia duboulayi*
Cycad, *Cycas armstrongii*
Death Adder, *Acanthophis rugosus*
Diamond Python, *Morelia spilota*
Dingo, *Canis lupus dingo*
Dunnart, *Sminthopsis murina*

Dusky Rat, *Rattus colletti*
European Toad, *Bufo bufo*
Feral Pig, *Sus scrofa*
Fijian Ground Frog, *Platymantis vitianus*
Fire Ant, *Solenopsis invicta*
Fox, *Vulpes vulpes*
Frenchie Beetle, *Lepidiota frenchi*
Freshwater Crocodile, *Crocodylus johnstoni*
Frillneck Lizard, *Chlamydosaurus kingii*
Fruit-bat, *Pteropus* species
Gamba Grass, *Andropogon gayanus*
Gastric-brooding Frog, *Rheobatrachus silus*
Giant Burrowing Frog, *Cyclorana australis*
Golden Toad, *Incilius periglenes*
Golden Tree Snake, *Dendrelaphis punctulatus*
Gray Squirrel, *Sciurus carolinensis*
Greater Bower Bird, *Chlamydera nuchalis*
Green and Golden Bell Frog, *Litoria aurea*
Green Tree Frog, *Litoria caerulea*
Grayback Beetle, *Dermolepida albohirtum*
Indian Mynah Bird, *Acridotheres tristis*
Ironwood Tree, *Erythrophleum chlorostachys*
Keelback Snake, *Tropidonophis mairii*
King Brown Snake, *Pseudechis australis*
Kiore (Pacific Rat), *Rattus exulans*
Koala, *Phascolarctos cinereus*

Komodo Dragon, *Varanus komodoensis*
Lace Monitor, *Varanus varius*
Land Mullet, *Bellatorias major*
Lion, *Panthera leo*
Long-necked Turtle, *Chelodina oblonga*
Lungworm, *Rhabdias pseudosphaerocephala*
Magnificent Tree Frog, *Litoria splendida*
Magpie, *Cracticus tibicen*
Magpie Goose, *Anseranas semipalmata*
Maple Tree, *Acer buergerianum*
Meat Ant, *Iridomyrmex purpureus*
Mongoose, *Herpestes javanicus*
Mother-of-Millions, *Bryophyllum delagoense*
Mountain Lion, *Puma concolor*
Northern Bluetongue Lizard, *Tiliqua scincoides intermedia*
Northern Quoll, *Dasyurus hallucatus*
Northern Spotted Gudgeon, *Mogurnda mogurnda*
Pandanus Palm, *Pandanus spiralis*
Pentastome (tongue-worm), *Raillietiella frenatus*
Phascogale, *Phascogale tapoatafa*
Pignose Turtle, *Carettochelys insculpta*
Pine Tree, *Pinus radiata*
Planigale, *Planigale maculata*
Platypus, *Ornithorhynchus anatinus*
Portuguese Millipede, *Ommatoiulus moreletii*
Prickly Pear, *Opuntia stricta*
Rabbit, *Oryctolagus cuniculus*
Rainbow Bee-eater, *Merops ornatus*

Rainbow Lorikeet, *Trichoglossus haematodus*
Red-bellied Black Snake, *Pseudechis porphyriacus*
Red Kangaroo, *Macropus rufus*
Rocket Frog, *Litoria nasuta*
Roth's Tree Frog, *Litoria rothii*
Saltwater Crocodile, *Crocodylus porosus*
Scotch Thistle, *Onopordum acanthium*
Scrub Turkey, *Alectura lathami*
Sea Eagle, *Haliaeetus leucogaster*
Short-necked Turtle, *Emydura tanybaraga*
Silver-leaved Paperbark Tree, *Melaleuca argentea*
Slaty-gray Snake, *Stegonotus cucullatus*
Snapping Turtle, *Elseya latisternum*
Speargrass, *Heteropogon triticeus*
Spinifex grass, *Triodia* species
Spotted-tail Quoll, *Dasyurus maculatus*
Starling, *Sturnus vulgaris*
Striated Pardalote, *Pardalotus striatus*
Stringybark Tree, *Eucalyptus tetrodonta*
Sugarcane, *Saccharum officinarum*
Sulphur-crested Cockatoo, *Cacatua galerita*
Taipan, *Oxyuranus scutellatus*
Tasmanian Devil, *Sarcophilus harrisii*
Tawny Frogmouth, *Podargus strigoides*
Thylacine (Tasmanian Tiger), *Thylacinus cynocephalus*
Tiger Salamander, *Ambystoma tigrinum*
Tongue-worm (Pentastome), *Raillietiella frenatus*
Water Buffalo, *Bubalis bubalis*
Water Dragon, *Intellagama lesueurii*
Water Hyacinth, *Eichornia crassipes*
Water Python, *Liasis fuscus*
Whistling Kite, *Haliastur sphenurus*
White-tailed Water Rat (Rakali), *Hydromys chrysogaster*
Willow Tree, *Salix* species
Wolf, *Canis lupus*
Woolly-butt Tree, *Eucalyptus miniata*
Yellow-spotted Goanna, *Varanus panoptes*

BIBLIOGRAPHY AND SUGGESTED READING

The following are just a few highlights from the huge literature on Cane Toads. I've focused on websites, books, and book chapters because they are relatively easy to find. I also include a selection of scientific papers for readers who wish to explore the original research.

WEBSITES

The Internet is full of information, and it's also full of pseudo-information. The problem is how to tell the evidence-based from the nonsense. My own toad research is mostly described on two websites, one that I organize myself (although my wife, Terri, does most of the work) and one that is run by the University of Sydney. The private website (www.canetoadsinoz.com) provides information designed for the general reader and has photographs of the people and projects I've talked about in this book. The Uni website (http://sydney.edu.au/science/biology/shine/) is only slightly more technical, and its Cane Toad section (. . . canetoad_research/) has lists of all our scientific papers, as well as summaries of them (in plain English). Plus, it tells you how to get hold of any of those papers (for free) if you want them. There's also a link to a downloadable booklet written to explain our research to children.

Other useful websites include those run by government departments, including the Western Australian Department of Parks and Wildlife (www.dpaw.wa.gov.au/plants-and-animals/animals/cane-toads) and similar bodies in Queensland (www.daff.qld.gov.au/__data/assets/pdf_file/0005/77360/IPA-Cane-Toad-PA21.pdf) and New South Wales (www.environment.nsw.gov.au/pestsweeds/CaneToads.htm). Also, the federal government set up a website for their "threat abatement plan" dealing with Cane Toads (www.environment.gov.au/resource/threat-abatement-plan-biological-effects-including-lethal-toxic-ingestion-caused-cane-toads).

The toad-busting community groups put a lot of effort into their websites. I am skeptical about the claims that are made on these sites, but they give you a good feel for people's motivations and passions. Because the sites have older as well as more recent newsletters, you can chart the changes in their hopes and claims as the wave of toads just kept on coming. The main ones for tropical Australia are those of the Stop The Toad Foundation (www.stopthetoad.org.au), Frogwatch NT (www.frogwatch.org.au), and the Kimberley Toad Busters (www.canetoads.com.au). These sites also provide information about how to tell a toad from a frog.

BOOKS

Although they were published before much of the research in Australia was conducted, the biology of the Cane Toad is discussed in these two books:

Kraus, F. 2009. *Alien Reptiles and Amphibians: A Scientific Compendium and Analysis*. Dordrecht, The Netherlands: Springer.
Lever, C. 2001. *The Cane Toad: The History and Ecology of a Successful Colonist*. Otley, UK: Westbury.

For an entertaining account of the toad's introduction to Australia, I recommend this book:

Turvey, N. 2013. *Cane Toads: A Tale of Sugar, Politics and Flawed Science*. Sydney: University of Sydney Press.

CHAPTERS IN BOOKS

These chapters summarize aspects of toad research:

Shine, R. 2010. Cane toad *(Rhinella marina)*. Pp. 299–310 in *Handbook of Global Freshwater Invasive Species* (R. A. Francis, ed.). Abingdon, UK: Earthscan.
Shine, R. 2014. The ecological, evolutionary and social impact of invasive cane toads in Australia. Pp. 23–43 in *Invasive Species in a Globalized World* (R. Keller and M. Cadotte, eds.). Chicago: University of Chicago Press.
Shine, R., and B. L. Phillips. 2014. Unwelcome and unpredictable: The sorry saga of cane toads in Australia. Pp. 83–104 in *Austral Ark* (A. Stow, ed.). Cambridge, UK: Cambridge University Press.

And here is Australian author Mark Dapin's account of the community groups' fight against toad invasion, which I quote in chapter 7:

Dapin, M. 2008. Who you gonna call? Toadbusters. Pp. 127–140 in *Strange Country*. Macmillan Australia, Sydney.

MAGAZINE ARTICLES

Phillips, B. 2016. Genetic "backburning" can stop cane toads. *Australasian Science* 37:14–17.
Shine, R. 2007. Toad kill. *Australasian Science* 28:16–20.
Shine, R. 2009. Controlling cane toads ecologically. *Australasian Science* 30:20–23.
Shine, R. 2011. It's evolution, but not as we know it. *Australasian Science* 32:16–19.

SCIENTIFIC PAPERS

If you're not a scientist, don't be put off by the formality of scientific publications. With a bit of effort you can get a lot out of them, even if you have no scientific training. Here are some that are pretty accessible (ordered chronologically by publication year). Check out my university website (details above) for a complete list of my papers on Cane Toads.

1975

Covacevich, J., and M. Archer. The distribution of the cane toad, *Bufo marinus*, in Australia and its effects on indigenous vertebrates. *Memoirs of the Queensland Museum* 17:305–310.

1986

Delvinquier, B. L. J. *Myxidium immersum* (Protozoa: Myxosporea) of the cane toad, *Bufo marinus*, in Australian Anura, with synopsis of the genus in amphibians. *Australian Journal of Ecology* 34:843–853.

1997

Barton, D. P. Introduced animals and their parasites: The cane toad, *Bufo marinus*, in Australia. *Austral Ecology* 22:316–324.

1998

Slade, R. W., and C. Moritz. Phylogeography of *Bufo marinus* from its natural and introduced ranges. *Proceedings of the Royal Society B* 265:769–777.

1999

Catling, P. C., A. Hertog, R. J. Burt, R. I. Forrester, and J. C. Wombey. The short-term effect of cane toads *(Bufo marinus)* on native fauna in the gulf country of the Northern Territory. *Wildlife Research* 26:161–185.

2003

Watson, M., and J. Woinarski. A preliminary assessment of impacts of cane toads on terrestrial vertebrate fauna in Kakadu National Park. Unpublished report, February.

2004

Boland, C. R. J. Introduced cane toads *Bufo marinus* are active nest predators and competitors of rainbow bee-eaters *Merops ornatus:* Observational and experimental evidence. *Biological Conservation* 120:53–62.

Phillips, B., and R. Shine. Adapting to an invasive species: Toxic cane toads induce morphological change in Australian snakes. *Proceedings of the National Academy of Sciences (USA)* 101:17150–17155.

Seton, K. A., and J. J. Bradley. 'When you have no law you are nothing': Cane toads, social consequences and management issues. *The Asia Pacific Journal of Anthropology* 5:205–225.

2006

Doody, J. S., B. Green, R. Sims, D. Rhind, P. West, and D. Steer. Indirect impacts of invasive cane toads *(Bufo marinus)* on nest predation in pig-nosed turtles *(Carettochelys insculpta). Wildlife Research* 33:349–354.

2007

Brown, G. P., C. M. Shilton, B. L. Phillips, and R. Shine. Invasion, stress, and spinal arthritis in cane toads. *Proceedings of the National Academy of Sciences (USA)* 104:17698–17700.

2010

O'Donnell, S., J. K. Webb, and R. Shine. Conditioned taste aversion enhances the survival of an endangered predator imperiled by a toxic invader. *Journal of Applied Ecology* 47:558–565.

2011

Brown, G. P., C. Kelehear, and R. Shine. Effects of seasonal aridity on the ecology and behaviour of invasive cane toads *(Rhinella marina)* in the Australian wet-dry tropics. *Functional Ecology* 25:1339–1347.

Brown, G. P., B. L. Phillips, and R. Shine. The ecological impact of invasive cane toads on tropical snakes: Field data do not support predictions from laboratory studies. *Ecology* 92:422–431.

Florance, D., J. K. Webb, T. Dempster, M. R. Kearney, A. Worthing, and M. Letnic. Excluding access to invasion hubs can contain the spread of an invasive vertebrate. *Proceedings of the Royal Society B* 278:2900–2908.

Shine, R., G. P. Brown, and B. L. Phillips. An evolutionary process that assembles phenotypes through space rather than time. *Proceedings of the National Academy of Sciences (USA)* 108:5708–5711.

2012

Crossland, M. R., T. Haramura, A. A. Salim, R. J. Capon, and R. Shine. Exploiting intraspecific competitive mechanisms to control invasive cane toads *(Rhinella marina)*. *Proceedings of the Royal Society B* 279:3436–3442.

Crossland, M. R., and R. Shine. Embryonic exposure to conspecific chemicals suppresses cane toad growth and survival. *Biology Letters* 8:226–229.

Kelehear, C., E. Cabrera-Guzman, and R. Shine. Inadvertent consequences of community-based efforts to control invasive species. *Conservation Letters* 5:360–365.

Pizzatto, L., C. Kelehear, S. Dubey, D. Barton, and R. Shine. Host–parasite relationships during a biological invasion: 75 years post-invasion, cane toads and sympatric Australian frogs retain separate lungworm faunas. *Journal of Wildlife Diseases* 48:951–961.

2013

Brown, G. P., M. J. Greenlees, B. L. Phillips, and R. Shine. Road transect surveys do not reveal any consistent effects of a toxic invasive species on tropical reptiles. *Biological Invasions* 15:1005–1015.

Brown, G. P., B. Ujvari, T. Madsen, and R. Shine. Invader impact clarifies the roles of top-down and bottom-up effects on tropical snake populations. *Functional Ecology* 27:351–361.

Cabrera-Guzman, E., M. R. Crossland, and R. Shine. Competing tadpoles: Australian native frogs affect invasive cane toads *(Rhinella marina)* in natural waterbodies. *Austral Ecology* 38:896–904.

Caller, G., and C. Brown. Evolutionary responses to invasion: Cane toad sympatric fish show enhanced avoidance learning. *PLoS ONE* 8:e54909.

Lindstrom, T., G. P. Brown, S. A. Sisson, B. L. Phillips, and R. Shine. Rapid shifts in dispersal behavior on an expanding range edge. *Proceedings of the National Academy of Sciences (USA)* 110:13452–13456.

Narayan, E. J., J. F. Cockrem, and J. M. Hero. Sight of a predator induces a corticosterone stress response and generates fear in an amphibian. *PLoS ONE* 8:73564.

Nyquist, J. R. Making and breaking the invasive cane toad. Master's thesis, Department of Social Anthropology, University of Oslo.

Somaweera, R., R. Shine, J. Webb, T. Dempster, and M. Letnic. Why does vulnerability to toxic invasive cane toads vary among populations of Australian freshwater crocodiles? *Animal Conservation* 16:86–96.

Tingley, R., B. L. Phillips, M. Letnic, G. P. Brown, R. Shine, and S. Baird. Identifying optimal barriers to halt the invasion of cane toads *Rhinella marina* in northern Australia. *Journal of Applied Ecology* 50:129–137.

2014

Brown, G. P., B. L. Phillips, and R. Shine. The straight and narrow path: The evolution of straight-line dispersal at a cane-toad invasion front. *Proceedings of the Royal Society B* 281:20141385.

Llewelyn, J., L. Schwarzkopf, B. L. Phillips, and R. Shine. After the crash: How do predators adjust following the invasion of a novel toxic prey type? *Austral Ecology* 39:190–197.

McCann, S., M. J. Greenlees, D. Newell, and R. Shine. Rapid acclimation to cold allows the cane toad *(Rhinella marina)* to invade montane areas within its Australian range. *Functional Ecology* 28:1166–1174.

Shine, R. A review of ecological interactions between native frogs and invasive cane toads in Australia. *Austral Ecology* 39:1–16.

2015

Brown, G. B., B. L. Phillips, S. Dubey, and R. Shine. Invader immunology: Invasion history alters immune-system function in cane toads *(Rhinella marina)* in tropical Australia. *Ecology Letters* 18:57–65.

Clarke, G., M. Crossland, C. Shilton, and R. Shine. Chemical suppression of embryonic cane toads *(Rhinella marina)* by larval conspecifics. *Journal of Applied Ecology* 52:1547–1557.

Hudson, C. H., B. L. Phillips, G. P. Brown, and R. Shine. Virgins in the vanguard: Low reproductive frequency in invasion-front cane toads. *Biological Journal of the Linnean Society* 116:743–747.

Jolly, C. J., R. Shine, and M. J. Greenlees. The impact of invasive cane toads on native wildlife in southern Australia. *Ecology and Evolution* 5:3879–3894.

Nelson, F. B. L., G. P. Brown, C. Shilton, and R. Shine. Helpful invaders: Can cane toads reduce the parasite burdens of native frogs? *International Journal for Parasitology: Parasites and Wildlife* 4:295–300.

Rollins, L. A., M. F. Richardson, and R. Shine. A genetic perspective on rapid evolution in cane toads. *Molecular Ecology* 24:2264–2276.

Shine, R., J. Amiel, A. Munn, M. Stewart, A. Vyssotski, and J. Lesku. Is "cooling then freezing" a humane way to kill amphibians and reptiles? *Biology Open* 4:760–763.

2016

Brown, G. P., and R. Shine. Frogs in the spotlight: A 16-year survey of the abundance of native frogs and invasive cane toads on a floodplain in tropical Australia. *Ecology and Evolution* 6:4445–4457.

Clarke, G. S., M. R. Crossland, and R. Shine. Can we control the invasive cane toad using chemicals that have evolved under intraspecific competition? *Ecological Applications* 26:463–474.

Hudson, C. M., G. P. Brown, and R. Shine. It is lonely at the front: Contrasting evolutionary trajectories in male and female invaders. *Royal Society Open Science* 3:160687.

Pettit, L. J., M. J. Greenlees, and R. Shine. Is the enhanced dispersal rate seen at invasion fronts a behaviourally plastic response to encountering novel ecological conditions? *Biology Letters* 12:20160539.

Phillips, B. P., R. Tingley, and R. Shine. The genetic backburn: Using rapid evolution to halt invasions. *Proceedings of the Royal Society B* 283:20153037.

Ward-Fear, G., D. J. Pearson, G. P. Brown, Balanggarra Rangers, and R. Shine. Ecological immunisation: *In situ* training of free-ranging predatory lizards reduces their vulnerability to invasive toxic prey. *Biology Letters* 12:20150863.

Wijethunga, U., M. Greenlees, and R. Shine. Moving south: Effects of water temperatures on the larval development of invasive cane toads *(Rhinella marina)* in cool-temperate Australia. *Ecology and Evolution* 6:6993–7003.

2017

Finnerty, P., R. Shine, and G. P. Brown. The costs of parasite infection: Removing lungworms improves performance, growth and survival of cane toads. *Functional Ecology* 31. In press.

Gruber, J., G. P. Brown, M. Whiting, and R. Shine. Geographic divergence in dispersal-related behaviour in cane toads from range-front versus range-core populations in Australia. *Behavioral Ecology and Sociobiology* 71:38.

Lillywhite, H., R. Shine, E. Jacobsen, D. DeNardo, M. Gordon, C. Navas, T. Wang, R. Seymour, K. Storey, H. Heatwole, D. Heard, B. Brattstrom, and G. Burghardt. Anaesthesia and euthanasia of amphibians and reptiles used in scientific research: Should hypothermia and freezing be prohibited? *BioScience* 67:53–61.

Taylor, A., H. I. McCallum, G. Watson, and G. C. Grigg. Impact of cane toads on a community of Australian native frogs, determined by 10 years of automated identification and logging of calling behaviour. *Journal of Applied Ecology* 54. In press.

Tingley, R., G. Ward-Fear, M. J. Greenlees, L. Schwarzkopf, B. L. Phillips, G. Brown, S. Clulow, J. Webb, R. Capon, A. Sheppard, T. Strive, M. Tizard, and R. Shine. New weapons in the Toad Toolkit: A review of methods to control and mitigate the biodiversity impacts of invasive cane toads *(Rhinella marina)*. *Quarterly Review of Biology* 92:123–149.

INDEX

Aboriginal attitudes towards Cane Toads, 80, 81*box*, 133, 140*fig*.
abundance of Cane Toads: at invasion front, 124–125; in long-colonized areas, 128
accelerating invasion of Cane Toads, 48, 60–78, 209, 224. *See also* dispersal rate of Cane Toads
accidental translocation of Cane Toads, 29, 55, 142, 162
adaptation, 25, 113, 216; in response to Cane Toad invasion, 112–115, 121–122, 233
Adelaide River floodplain, 37, 41, 240–241
African Clawed Frog, 83–86
aggregation of tadpoles, 183
Agile Wallaby, 42
agriculture, 16–17, 38, 139, 155, 164
alarm pheromone, 191–193, 196–199; identify specific chemical, 192; no effect on native tadpoles, 192
Alford, Ross, 58, 123, 128, 179
ambush predation, 40, 58, 95, 102, 117, 119
American Bullfrog, 84
Amiel, Joshua, 206
amphibian invaders, 84–85
amphibians, impact of Cane Toad on, 34, 49, 90–94, 193
amplexus, 180–182, 181*fig*.
Andersen, Hans Christian, 135
animal-welfare activists, 212, 238
Antilopine Wallaroo, 39

anti-toad hysteria: strongest at invasion front, 132, 145
ants as predators of Cane Toads, 187*fig*., 188
Anura, 3, 7, 84
apex predators, 4, 118, 123, 222, 226, 232
appearance of Cane Toads, 4*fig*., 8fig., 11*fig*., 16–23, 27*fig*., 71*fig*., 119, 127*fig*., 181*fig*., 135, 179, 236; public revulsion, 228
Archer, Mike, 34
arid conditions, 31–32
arms race, 2, 165
arrival of Cane Toads at Fogg Dam, 36, 47, 54, 92
Asian frog-eating snakes, 120
Asian goannas, 126
Asian House Gecko, 87; as host for pentastome parasite, 88
attitudes of the public to Cane Toads, 10, 12, 23–24, 34, 80, 130–136, 139; acceptance, 228–230, 239; debates, 148; declining interest, 152; hysteria, 130, 133, 145, 151, 239; media focus, 204
attractant pheromone, 193–196, 199, 228; identify chemical, 195; risk to public, 204–206
Australiana, 131
Australian Academy of Science, 209
Australian government, 22, 23, 38, 48, 131, 151, 165–166, 178, 203, 219, 237

Australian invasion of the Cane Toad, 59*map*
Australian Research Council (ARC), 48, 153, 174
Australia's isolation though evolutionary time, 2
aversion learning, 108–115, 209–215, 219, 233; Bluetongue Lizards, 215; fishes, 50, 90; Freshwater Crocodiles, 215; goannas, 215–217; Northern Quolls, 50, 211–218. *See also* teacher toads

bacteria, 68, 77, 166
Baker, Paul, 158
Balanggarra Rangers, 217–220, 217*fig.*
Balzac, Honore de, 1
Bandicoot, 45–46, 239
ban on release of Cane Toads, 22
Barbados, 16, 26, 33
Barkley Tablelands, 24
Barramundi, 42, 88–90, 121, 168
Barton, Di, 168, 170
basic vs. applied science in pest control, 178
Beatrice Hill Farm, 48
Beckmann, Christa, 134, 138; research on toad impact on birds, 50, 104–105
bee-hives, impact of Cane Toads on, 24, 79
beetle larvae as sugarcane pests, 16–23, 119, 135, 236
beetles, 6, 43, 77, 199
behavioral flexibility: of Cane Toads, 28; vs. invasion success, 6
Bell, Arthur, 17–20, 235
biggest of the biggest, the, 7*box*
biodiversity crisis in Australia, 12, 120, 129, 139, 148, 154, 164, 224, 238–239
biological control, 1, 17–19, 82, 174
biotic resistance, 108–110, 128, 234, 242
bird roosts, 15
birds, impact of Cane Toads on, 33, 50, 103–105, 118, 125–126
Black Death, 1
Black Kite, 105
Black Rat, 55, 93, 136, 231
Black-spined Toad, 84, 134
Black Whipsnake, 43

Bluetongue Lizard, 100–101, 107, 110, 121, 215, 222, 230; geographic variation in vulnerability to toad toxin, 121–122; impact of toad invasion, 121
body plan of Cane Toads, 67
body size of Cane Toads: at invasion front, 123–124, 209, 214, 232; maximum, 7
body size vs. invasion success, 6
Boland, Chris, 104
Boulter, Sandy, 143
Bowcock, Haley, 182
breeding biology of Cane Toads, 5, 21, 24, 29–33, 57–58, 70–72, 94, 162–163, 179–183, 197–199, 209, 214; amplexus, 180; breeding pond preferences, 186–187, 204; in captivity, 61–63; female control, 182; male release call, 180–181; in their native range, 187; oviposition, 204; seasonal sex dimorphism, 180, 181*fig.*
Brett, Sarah, 139
Broome, Western Australia, 223
Brown, Greg, 40–43, 41*fig.*, 47*fig.*, 48–49, 63–64, 69, 76–80, 91, 100–102, 116–117, 124–126, 173, 175, 211, 219*fig.*, 229, 235, 240; long-term research projects, 150, 163; pet buffalo, 42; pet chickens, 75; pet dog, 62; pet fish, 89; pet lizard, 46; radio-tracking toads, 52, 54, 58–59, 65–66, 74; toad immunology, 63, 68. *See also* Shilton, Cathy
Brown Rat, 126–127
Brown Tree Snake, 120, 230–231
Bruning, Bas, 182
bucketful of Cane Toads, 11*fig.*
bufagenin, 9, 195
buffering the impact of Cane Toads, 203, 209, 213, 222, 226–227, 234, 237. *See also* aversion learning; conditioned taste aversion; taste aversion; teacher toads
Bufobabble, 145–150, 175–177, 206, 210, 235, 238
Bufo horribilis, 13
Bufonidae, 2, 57; ecology of, 3; invasion of, 5; reproduction of, 3
Bufonite, 197–201, 198*fig.*, 207, 228; dependence on egg size, 201; impact on breeding behavior, 200; impact on

native tadpoles, 200; use for toad control at invasion front, 207. *See also* suppression pheromone
bufotoxins, 9, 37, 128
Bulcock, Frank, 22–23
bureaucracy, 205, 212
Bureau of Sugar Experiment Stations, 18, 20*fig.*, 56, 86, 235–236
Burton's Legless Lizard, 99
bush tucker, 81, 95
butterflies, 69, 77
Buzacott, James, 20, 236
by-catch in toad traps, 158

Cabrera-Guzmán, Elisa, 126, 128, 188; interactions between toads and insects, 50
Cactus Moth, 18
caecilians, 3
Cairns, Queensland, 20, 26–27, 60–61, 88, 123, 132, 207, 266
Camel, 55–56
Campa tribe of Peru, 9
cane beetle, 16–23, 119, 135, 236
Cane Toad equation, 77
Cane Toad photographs, 4*fig.*, 8*fig.*, 11*fig.*, 27*fig.*, 71*fig.*, 127*fig.*, 181*fig.*
Cane Toad races, 132
"Cane Toad's Plain Code" (song), 132
Cane Toad Wars, 89, 150, 228, 238–242
cannibalism in Cane Toads: consumption of eggs, 194, 201; consumption of metamorphs, 128, 148–149, 184; diel timing, 185; toe-luring, 184
Capon, Rob, 9, 36, 192, 195
Carp, 55
Carr, Caleb, 36
cat, 1, 29, 34, 55–56, 83, 93, 106, 108, 133, 158, 189, 190, 231
Catling, Peter, 91
changing rhetoric from community groups, 48, 145–152
Charles, Prince, 131
Charles Darwin University, 93
Chaunus marinus, 15
chemical ecology of tadpoles, 189. *See also* alarm pheromone; attractant pheromone; suppression pheromone

Chestnut Blight Fungus, 1
Chicken Little, 150
Chimpanzee, 166
Christian, Keith, 93
Chytrid Fungus, 1, 85, 86, 179
citizen science, 154
Clarke, Greg, 200
Clarke, Rachael, 131
classification of the Cane Toad, 13–15; scientific description, 13; scientific name, 13. *See also* common name of the Cane Toad
classification of toads, 3*box*
climate matching, 30
climatic challenges encountered by Cane Toads in Australia, 26, 222
clutch size of Cane Toads, 21, 60, 155, 181–182
Cockatoo, Sulphur-crested, 2, 43, 190
coevolution, 2, 165
cold tolerance of Cane Toads, 30–32
collaboration: between researchers and indigenous rangers, 217–218; between researchers and private enterprise, 207
collateral impacts of toad control, 187, 192, 200
commercial utilization of Cane Toads, 160
common name of the Cane Toad, 14
Commonwealth Scientific and Industrial Research Organisation (CSIRO), 22, 91, 164–167, 178–179
communication in tadpoles, 183, 191. *See also* alarm pheromone; attractant pheromone; suppression pheromone
community groups, 12. *See also* Frogwatch NT; Kimberley Toad Busters; Stop The Toad Foundation; toad-busters
competition between Cane Toads and frogs, 184; as adults, 29–30, 50, 92, 190*box*, 191, 232; as tadpoles, 188–191
competition between Cane Toads and other toad species, 30
complexity of impacts of Cane Toads, 122, 233
conditioned taste aversion, 219; in Blue-tongue Lizards, 215; in fish, 110; in frogs, 110; in goannas, 112, 217; in Planigales, 109; in quolls, 211; in snakes, 112. *See also* teacher toads

conflict: between scientists and animal-welfare groups, 212, 238; between scientists and community groups, 148–150, 175–177, 206, 210, 235, 238
conflicts among community groups, 159, 164, 178, 235
conservation priorities, 232, 237, 239–242
controlling Cane Toads, 145, 155, 179, 228; use of Bufonite, 197–207; use of native fauna, 204. *See also* fences to exclude toads; trapping adult Cane Toads; trapping Cane Toad tadpoles
Coqui, 84–85, 239
Corellas, 43
Corroboree Frog, 83
corticosterone, 32
Costa Rica, 14, 83
Covacevich, Jeanette, 34
Cox, Eric, 52, 62
creationists, 25, 69
Cremona, Teigen, 214
Crimson-spotted Rainbowfish, 90
crocodile conservation, 95–97
Crossland, Michael, 41*fig.*, 51–52, 62, 63, 179, 189–207, 219*fig.*
CSIRO, 22, 164–167, 178–179
culling adult toads, 155, 156, 160, 163, 232
Cunnamulla, Queensland, 161
curtailing further spread of Cane Toads, 203, 222–224

Dapin, Mark, 136, 141, 252
Darwin, Charles, 25, 27, 48, 65, 69, 71–75, 113, 181
Darwin (city), Northern Territory, 7, 11, 26, 28, 37, 48, 52, 56, 59–60, 62, 67, 88–91, 96–97, 105, 122, 130, 137–140, 143, 145, 152, 158–159, 161, 164, 175, 205, 210–211, 229
Death Adder, 34, 44, 53, 102, 112–118
decline in abundance of Cane Toads post-invasion, 124–125
decline of toad-busting groups, 151, 236
Delvinquier, Benoit, 86–87
Department of Parks and Wildlife (DPaW): Western Australia, 172, 217–218, 237

deranged ideas about controlling toads, 142*box*
DeVore, Jayna, 200
Dexter, Raquel, 16–17
Diana, Lady, 131
diet breadth vs. invasion success, 6
diet of Cane Toads, 7, 8*fig.*
different, but not inferior, 96–97*box*
Dingo, 231
direct impact of Cane Toads on wildlife, 116–119
dispersal of Cane Toads, reliance on moist conditions, 222
dispersal rate of Cane Toads, 56–58; effect of lungworms, 173; at the invasion front, 123; path straightness, 64
domestic pets: impact of Cane toads on, 30, 83; impact on chickens, 134; impact on dogs, 133
Doody, Sean, 118
DPaW, 172, 217–218, 237
dryland corridor, 223–226
Dubey, Sylvain, 171
Ducatez, Simon, 200
dung beetles, 28, 52, 79–80
Dunnart eating toad, 111*fig.*
Dusky Rat, 126
Dusty, Slim, 132

Easteal, Simon, 57
ecological catastrophe, 108, 120, 131, 165, 208, 235, 240
ecological flexibility of Cane Toads: ability to exploit disturbed habitats, 93; artificial lights, 28, 93; cities, 31; high cold regions, 32; roads, 28
ecological impact of Cane Toads, 4, 33, 49, 80, 115, 120. *See also* poisoning
ecological traps, 186; breeding site selection, 186; egg size reduction, 201; seasonal timing of reproduction, 187; vulnerable to insect predators, 188; vulnerable to native tadpoles, 188
ecology of Cane Toads, 6; ecological generalists, 185; time of activity, 32
ecosystem engineer, 15
ectothermy, 96–97*box*

eggs of Cane Toads, 50, 155, 156*fig.*, 185; geographic variation in egg size, 201–202; tadpoles eat eggs, 194
Einstein, Albert, 68, 155, 190
Elapidae, 37, 112–113
El Niño, 27
Endler, John, 76
enemy release, 169
environmental change, 28
eradication of Cane Toads, 159–163, 176–177, 186, 196–197, 208, 222, 229, 232
ethical issues with experiments on animals, 119, 205–206, 212–213, 238
ethical regulation of animal research, 212–213*box*
Ethics Committee regulations, 205–206, 213*box*
ethics of keeping wildlife as pets, 190*box*
European Toad, 22, 56, 181
euthanasia of Cane Toads, 147, 205–206
evolution, 2, 7, 8, 12, 25–27, 36–37, 60–61, 65–78, 165, 179, 181, 183, 200–203, 208–209, 218, 224, 233
evolutionary rescue, 112–115, 121–122, 233
evolutionary trade-offs, 26
evolution of rapid dispersal, 58–78, 123, 207–209, 224–225, 233
evolution of resistance to toad toxin in predators, 97, 112–113, 120–126, 216
evolution of toad-smart behavior in predators, 112–114, 120, 208
exploitation of Cane Toads by native species, 116–118, 123, 125–129, 164, 232
extinction of amphibians, 85
extirpation of Cane Toad populations: Cunnamulla, 161; Darwin, 161; Port Macquarie, 161; St George, 161; Taren Point, 177

farm dams, 28, 194, 197, 222–223
fear of Cane Toads, 80, 130
fecundity of Cane Toads, 21, 155, 181–182
fecundity vs. invasion success, 8
feeding in Cane Toads, 7, 8*fig.*
fences to exclude toads, 142, 159–160; effective spatial scale, 159–160; impact on native wildlife, 159
Fenner, Frank, 165, 167

Fiji, 29, 82, 124
Fijian Ground Frog, 82
financial incentives for toad control: bounties, 160, 178; fertilizer, 160; leather, 160
Finnerty, Patt, 173–174
Fire Ant, 76
firearms for toad control, 142, 160
fishes, aversion learning in, 50, 90
Florance, Dan, 159
Florida, 29, 82–83, 85
Fogg Dam, 36–46, 39*fig.*, 91–120, 150, 163–164, 187, 193–196, 211, 215, 230, 234, 240–242
foraging biology of Cane Toads, 7, 8*fig.*
Fourth Congress of the International Society for Sugar Cane Technologists, 16
Fox, 12, 55–58, 108, 123, 155, 231
Franklin, Michelle, 61
Freeland, Bill, 47, 91–92
French Guiana, 13, 169
Frenchie Beetle, 18
Freshwater Crocodile, 4, 39*fig.*, 45, 98*fig.*, 97–98, 107, 110, 118, 158; aversion learning in, 215
Frillneck Lizard, 35, 42, 99, 105
Froggat, Walter, 21–23, 33, 235–236
Froggatt Prize for Science, 236
frogs, impact of Cane Toads on, 34, 49, 90–94, 193
Frogwatch NT, 91, 139, 140–143, 148, 152, 159–160, 207, 210
fruit-bat, 166

Gamba Grass, 55
Gastric-brooding Frog, 83
genes, 25, 166–167, 171, 178, 194, 207, 216, 218, 224–226; for accelerated dispersal, 65, 70–74
genetically-engineered virus, 142, 165–167, 178
genetic back-burn, 207, 225–228
geographic distribution of Cane Toads, 30–31, 59*map*
geographic variation in Cane Toad traits: abundance, 124–125; body size, 123–124; egg size, 201–202; parasite load, 125

geographic variation in impact of Cane Toads: Bluetongue Lizards, 121; in crocodiles, 98; Red-bellied Blacksnakes, 113
Georges, Arthur, 95
Giant Burrowing Frog, 92
Giant Toad, 14
Gleitzman, Morris, 131
global spread of Cane Toads, 16, 82
goanna, 118–120, 220, 217*fig*.; aversion learning, 216–217. *See also* varanid lizards; Yellow-spotted Monitor
Golden Toad, 83
Golden Tree Snake, 46, 120
González-Bernal, Edna, 79
Goodgame, Dean, 141
Grahame, Kenneth, 57, 155
Grayback Beetle, 18
Gray Squirrel, invasion of Italy, 239
Greater Bower Bird, 43
Green and Golden Bell Frog, 234
Greenlees, Matthew, 49–50, 91–92, 121, 126, 163
Green Tree Frog, 93, 138–139, 172, 189–190; tadpoles compete with toad tadpoles, 189
Grigg, Gordon, 57, 91
Groffen, Jordy, 175–177
Gulf of Carpentaria, 26

habitat use of Cane Toads: generalist, 15, 28, 60, 93, 184–185; metamorphs restricted to water's edge, 128
Hagman, Mattias, 186, 191, 195
Hamilton Island, 163
Hamley, Tim, 95
Haramura, Takashi, 196–197
Hartigan, Ashlie, 87
Hawaii, 5–6, 14–19, 21–22, 33–34, 57, 70, 85–86, 125, 168–169, 171, 179, 200, 224, 239
Hawke, Bob, 165
Hearnden, Mark, 179
heritability of dispersal traits, 61–64, 225
hermaphrodite, 168
history of Cane Toad invasion, 59*map*, 169; breeding and release in Australia,

21; public welcome to Australia, 21; shipment to Hawaii, 17
Hogben, Lancelot, 85
host-parasite interactions, 2, 83–89, 120, 125, 167–177
hotline to report Cane Toads, 137
how deadly are Cane Toads?, 9*box*
how to kill a Cane Toad humanely, 205–206*box*
humane euthanasia of Cane Toads, 205–206*box*
human population growth, 234
Hyatt, Alex, 167

identification of Cane Toads, 137–138*box*, 229
immune system compromised at Cane Toad invasion front, 68, 72, 123
impact of Cane Toad control in the range core, 232
impact of Cane Toad control on native wildlife: attractant pheromone, 195; suppression pheromone, 200; toad exclusion fences, 159
impact of Cane Toads in other countries: Hawaii, 34; Puerto Rico, 34
impact of Cane Toads on wildlife: on Brown Tree Snakes, 120; on crocodiles, 98; on Death Adders, 102, 114; on fish, 89; on Freshwater Crocodiles, 107, 110; on goannas, 100–111; on Golden Tree Snakes, 120; on insects, 33, 35, 79, 80, 82; on lizards, 101, 107, 110; on quolls, 209; on snakes, 100, 116, 209; on turtles, 94, 95
impact of toad-busting, 148–152, 163–164
importation of Cane Toads to Australia, 20*fig*.
Indian Mynah Bird, 55
indigenous attitudes towards Cane Toads, 80, 81*box*, 133, 140*fig*.
indirect impacts of Cane Toads, 116–120
insects, impact of toads on, 17, 33, 58, 232
Insular Experiment Station, Puerto Rico, 16
integrated approach to pest control, 226
Interfet Frog, the, 134*box*
international translocation of toads, 16–17, 29

introduction of Cane Toads to Australia, 18–23
intuition vs. evidence-based conservation, 232–239
invasion front populations of Cane Toads, 58, 60, 65–67, 71*fig.*; escape from parasites, 172; low reproductive rate, 123; rapid dispersal, 123, 224; reduced survival, 74
invasive animals in Australia: Black Rats, 55, 231; camels, 55; Carp, 55; feral cats, 55, 108, 123; foxes, 55, 108, 123; horses, 239; Indian Mynah Birds, 55; Portuguese Millipede, 125; Rabbits, 55, 108, 123; Water Buffaloes, 55, 231
invasive plants in Australia: Gamba Grass, 55; Maple Trees, 55; Mother-of-Millions, 121; Scotch Thistles, 55; Willow Trees, 55
invasive species, defining: Brown Tree Snake, 230; Dingo, 231; Platypus, 230
Ironwood Tree, 42
Ishigaki Island, Japan, 82, 160

Jaguar, 142
Jamaica, 16
Japan, 82, 160
Jolly, Chris, 101, 111, 118
Judd, Charles, 125

Kakadu National Park, 34, 47–48, 54, 58, 95, 99, 164, 167, 186, 214, 231–233, 240
Kearney, Mike, 31
Keelback Snake, 44–46, 126
Kelehear, Crystal, 88, 170–174, 211
Key Threatening Process, 203
killing Cane Toads, 147, 205–206
Kimberley, Western Australia, 26, 139–144, 151, 159, 173–174, 177, 179, 216, 220–223, 226
Kimberley Toad Busters (KTB), 138–140, 145, 214; analysis by anthropologist, 145; conflicts with other groups, 143, 148–153, 159, 175–177, 210; government funding, 143; inflated claims, 146–147; lungworm research, 174–176, 226; military terminology, 147; social

inclusion policy, 141; transfer of lungworms, 176
King Brown Snake, 99, 102, 103*fig.*, 112, 118, 230–232, 240
Kiore, 108
Koala, 2, 11, 149, 231
Komodo Dragon, 111
KTB. *See* Kimberley Toad Busters
Kununurra, Western Australia, 26–27, 139–147, 157, 164, 175, 207, 210, 215

Lace Monitor, 101, 111–112
Land Mullet, 101
landscape-scale control of Cane Toads, 164, 200, 218, 227
larval ecology of Cane Toads, 50–51; aggregation, 182; aquatic, 185; consume conspecific eggs, 194
legislation to protect Cane Toads, 23
leg length of Cane Toads at invasion front, 66–67, 123
Lesku, John, 206
lethal toxic ingestion, 33. *See also* poisoning
Letnic, Mike, 159
Le Tour de Cane Toad, 131
Lettoof, Damian, 119
Levey, Ken, 48
Lewis, Mark, 131
life history of Cane Toads: aquatic eggs, 185; fecundity, 183; high mortality rates, 185; metamorphosis, 182
Lindner, Dave, 99
Linnaeus, 13, 14
lion, 99, 190, 231
Little Mulgrave River, Queensland, 21, 226
Little Red Riding Hood, 210, 214
Lizard Island, Queensland, 115
Llewelyn, John, 114–115
Long-necked Turtle, 44, 94–95
Longreach, Queensland, 26; Cane Toad at Longreach, 27*fig.*, 71*fig.*
low-risk approach to buffering Cane Toad impact, 210
lungworm effect on Cane Toads: dispersal rate, 173; low blood cell counts, 168; manipulate host behavior, 173–174
lungworm effect on frogs, 168–173

lungworm research: by community group, 175–177; by Team Bufo, 170–174
lungworms, 88–89, 119–120, 179, 226; absence from toad invasion front, 173; evolution of larger eggs, 173; geographic origin, 169; life cycle, 168; use for controlling toads, 168–177, 169*fig.*
Lyons, Joseph, 22

MacFarlane Burnett Medal, 209
Macquarie University, 90
Madagascar, 121–122
Madsen, Thomas, 38–39, 102, 125–126
Magnificent Tree Frog, 172–174
Magpie, 104
Magpie Goose, 43, 139
male-male rivalry in Cane Toads, 180; seasonal elaboration of musculature in males, 180
Marine Toad, 14
mark-recapture studies on snakes at Fogg Dam, 39, 44–46
marsupials, aversion learning in, 50, 211, 215, 218
Martinique, 16
mathematical models of toad control, 149
measuring ecological impact of an invasion, 35, 79–129
Meat Ant, 127–129, 188*fig.*, 226
mechanisms of invader impact, 116; complexity of, 228, 232
media stories about Cane Toads, 114, 148, 151, 204; Black-spined Toad, 134; long-legged toads, 67; toad research, 204
Meringa, Queensland, 20–22
metamorph Cane Toads, 50, 94; cannibalism, 194; predation by ants, 204; restriction to edge of pond, 128, 182–185
Middle Point, 38–43, 56, 61–64, 93, 109, 116, 176, 181, 192
Middle Point Village School, 39, 62
military analogy to fighting pest species, 60, 136, 142, 147, 178
mobility vs. invasion success, 8
mongoose, unaffected by toad toxins, 125
Moritz, Craig, 166

mortality, effect on population viability, 119, 241
Mother-of-Millions, 121–122
motivating toad-culling: bounties, 160, 178; fertilizer, 160; leather, 160
Mungomery, Reg, 5, 18–21, 23, 26, 29, 33, 56, 64, 86, 135, 142, 177, 226–227, 233–236; office 20*fig.*
Mungomery Library, 64
Munn, Adam, 206
mutation, 25, 31, 65, 77
mutation surfing, 74
myths about toads, 11, 145–150, 175–177, 206, 210, 235, 238
Myxidium, 86
Myxosporeans, 86–87

name of the toad, the, 14*box*
Narayan, Ed, 82
native range of Cane Toad, 13, 82, 130
native wildlife adapting to toads, 108–110, 208, 233
natural choke point, 223
natural selection, 25, 31, 36, 69–77, 112–114, 117, 194, 233
Nature (scientific journal), 66
Natusch, Dan, 124
nausea-inducing chemical, 108, 211, 215
negative impacts of Cane Toads, 119, 123. *See also* impacts of Cane Toads; poisoning;
Nelson, David, 50, 90
New Caledonia, 234
New South Wales government, 24, 218–219, 251
New South Wales Scientist of the Year, 218
New York Times (newspaper), 66
New Zealand, 29, 55, 79, 108, 234
niche modeling, 31
Nonsuch Island, Bermuda, 163
Northern Bluetongue Lizard, 100–101, 107, 110, 121, 215, 222, 230; geographic variation in vulnerability to toad toxin, 121–122; impact of toad invasion, 121
northern New South Wales, 65, 163
Northern Quoll, 4, 106–107, 110, 114–115, 211–218, 222, 231
Northern Spotted Gudgeon, 90

Northern Territory, 7, 9, 24–26, 30, 37–38, 40, 47, 60–61, 66, 88–89, 91, 93, 95, 98–99, 119, 121, 131, 136, 139–141, 151, 158, 179, 201, 207
Northern Territory government, 38, 47, 52, 158
Northern Territory invaded by Cane Toads, 25
NT News (newspaper), 67, 130, 133, 159
Nyquist, Jon Rasmus, 139, 145

O'Donnell, Stephanie, 50, 211, 215, 218
Old Testament, 133–134
Olympic Village effect, 70
Oombulgurri, Western Australia, 216–217
Order of Australia, 52
Origin of Species, The (book), 48, 69
other invasive amphibians, 84–85*box*
oviposition in Cane Toads, 204

Pandanus Palm, 43–44, 46
Papua New Guinea, 16, 82, 230
parasite lag, 172
parasite manipulation of their hosts, 174
parasites for Cane Toad control, 167–177
parasites of Cane Toads, 83–88, 120, 167–177; absence from invasion front, 125; escape from native-range parasites, 169
parasitologists, a scarcity of, 170*box*
parotoid glands, 8–10, 196, 204
patent for tadpole trapping, 207
path straightness, 58, 64
Peacock, Tony, 144
Pearson, David, 237
peculiar ideas about toad control: free beer, 178; helicopter with vacuum cleaner, 142; introduce jaguars, 142; microwave arrays on fighter jets, 142
Pemberton, Cyril, 6, 17–18, 22
pentasomes, 87–88
personality of lizards, 220
pheromones: alarm pheromone, 191–193, 196–199; attractant pheromone, 193–196, 199, 204–206, 228; suppression pheromone, 197, 198*fig.*, 200–201, 207, 228
Phillips, Ben, 37, 49, 53–54, 62–64, 69, 72, 76–78, 113–116, 121–124, 181, 223–225,

235; 41*fig.*, 219*fig.*; research on evolution of toad dispersal, 60–66, 72, 77; research on toad impact, 101–102
physiological resistance to toad toxins, 97, 112–113, 120–126, 216
physiology of Cane Toads: endocrine, 32; immunology, 68, 123
Pignose Turtle, 95, 118
pigs, 29, 95, 144, 155, 175
Pilbara barrier, 223–227
pine trees, 77
Pizzatto, Ligia: research on cannibal toads, 148; research on lungworms, 171–172
plague of frogs, 134
Planigales, 109–110, 211
Platypus, 2, 230
poisoning of domestic pets, 30, 83; of chickens, 134; of dogs, 133
poisoning of humans, 9, 19
poisoning of wildlife, 107, 164, 208–209, 232; of birds, 33, 103–105; of crocodiles, 95, 107, 110; of fish, 89; of lizards, 99–101 107, 110, 209; of quolls, 107, 110, 209–211; of snakes, 102, 101, 116, 209; of tadpoles, 193; of turtles, 94–95
poison used to kill juvenile toads, 160
political alliances of community groups, 144
political context of conservation wars, 12, 17–24, 134, 143–145, 151–154, 176–179, 210, 237–238
political context of toad research, 12, 61, 148, 154, 167, 216, 236–242
political dimensions of animal ethics, 212
politics, interstate rivalry, 14, 22, 24
population structure of toads behind invasion front, 214
Port Macquarie, New South Wales, 161
Portuguese Millipede, 125
positive impacts of Cane Toads, 116–119, 125; on ants, 128–129; on rats, 126–127; on small reptiles, 232–234; on snakes, 116–117; on spiders, 128; on turtles, 95; on water beetles and bugs, 128–129
positive impacts of community groups, 236
predation impact on prey abundance, 119

predation on Cane Toads: by ants, 128–129; by rats, 126–127, 127*fig.*; by spiders, 128; by water beetles and bugs, 128–129
predator body size determines vulnerability, 232
predicting ecological impact, 30–35, 120, 235
predicting geographic distributions, 24, 30–33, 59*map*
pregnancy test, 62, 84–86
Price-Rees, Sam, 50, 101, 215, 218
Prickly Pear, 18, 33
Prime Minister's Prize for Science, 218–219
Proceedings of the National Academy of Sciences USA (scientific journal), 76
Public Enemy Number One, 81, 123, 139, 218
public perception of ecological impact of Cane Toads, 34, 48–49, 136, 141
public profile of Cane Toads, 10, 12, 23–24, 34, 80, 130–136, 139; acceptance, 228–230, 239; debates, 148; declining interest, 152; hysteria, 130, 133, 145, 151, 239; media focus, 204
public status of scientists declining, 237
public welcome of Cane Toads to Australia, 20–21
Puerto Rico, 14, 16, 18, 21, 34, 169

quarantine, 22, 86, 169, 171, 242
Queensland, 79, 83, 103–106, 114, 121, 124, 128, 130–135, 137, 161, 163, 169–170, 179, 191–192, 200–201, 207, 215, 222, 224–225, 230, 233
Queensland Bureau of Sugar Experiment Stations, 18, 20*fig.*
Queensland Government support for toad introduction, 17–19, 22
quolls, 34, 106–107, 110, 114–115, 211–218, 222, 230–231

Rabbit, 12, 33, 51, 55, 58, 108, 123, 165
radio-tracking, 8, 52, 59, 61, 66, 101, 119, 126, 127*fig.*, 162, 173, 215–217
Rainbow Bee-eater, 43, 104
rainforest, 26
Rakali, 46, 126

Ramos, Menendez, 16
Rana marina, 14
range edge, 37, 72–77, 201, 224
rapid evolution, 30–31, 61–66, 113, 230–233
rapid evolution in Cane Toads, 67, 70–78, 120–126, 202, 233
rate of invasion of Cane Toads, 48, 57–78, 209, 224
rat-killed Cane Toad, 127*fig.*
Raven, Milly, 182
Red-bellied Black Snake, 2, 105, 113–114, 121
Reeves, Matthew, 133
regulations assist Cane Toad spread, 29
Reimer, William, 82
reproduction in Cane Toads: female inability to call, 180; male release call, 180–181; male reproductive season, 181; male rivalry, 180
reproductive output of Cane Toads: clutch size, 155, 182; number of clutches per year, 181
reptiles, impact of Cane Toads on, 34, 101, 111, 118. *See also* poisoning
research by community groups, 152–153
resilience of Australian wildlife, 114, 233–234, 243
Rhinella marina, 15
rice farming, 37–38
Riesenzellen, 191
risk to public from toad-control measures, 204–206
road-kill, 8*fig.*, 9–10, 105, 115, 117, 119, 131, 171
Rocket Frog, 3, 46
Rostand, Jean, 228
Roth's Tree Frog, 93

Sabath, Michael, 58
salamanders, 3, 84–85
Salim, Angela, 195
salinity, 6, 14
Saltwater Crocodile, 40, 45, 94–95, 97–98
Sarina, Queensland, 20
sausage, 215–218
Sawyer, Graeme, 130, 140–142, 152, 158–160, 210, 218; advocated firearms for toad control, 160; conflicts with

scientists, 150; leader of Frogwatch NT, 140–142, 210; Lord Mayor, 145
scapegoat, 135, 139
Schwarzkopf, Lin, 58, 159
Scotch Thistle, 55
Scott-Virtue, Lee, 140–143; married on a toad-bust, 141
Scrub Turkey, 118
Seabrook, Wendy, 57
Sea Eagle, 94
seasonal inactivity of native frogs, 27, 93
seasonal timing of reproduction of Cane Toads, 29, 71*fig.*, 128, 180–188
Seba, Albertus, 14
Semeniuk, Mark, 186
Seuss, Dr., 55
sex differences in Cane Toads: aquatic eggs, 185; body size, 179; calling behavior, 180; color, 179–180; skin rugosity, 179
sexual behavior of Cane Toads, 179–185, 181*fig.*
Shakespeare, William, 108, 135
Shanghai Daily (newspaper), 66
Shilton, Cathy, 67–68
Shine, John, 209
Shine, Rick, 41fig., 218*fig.;* early years, 53
Short-necked Turtle, 95
Silver-leaved Paperbark Tree, 43, 230
Slade, Rob, 166
Slaty-Gray Snake, 46
snake research at Fogg Dam, 39, 40–46, 102
snakes, impact of Cane Toads on, 34–35, 101. *See also* poisoning
Snapping Turtle, 95
Sniffer dogs, 156–157
social behavior of Cane Toads, 7; tadpole aggregation, 182
societal context for conservation, 231–240
Somaweera, Nilu, 137
Somaweera, Ruchira, 50, 98–99, 137
spatial sorting, 68–78, 233; in bacteria, beetles, butterflies, trees, starlings, vines, voles, 77, 224, 233; mathematical models, 72, 77; oarsmen analogy, 73
Speargrass, 241

speed of invasion, 25, 48, 56–78, 209, 224. *See also* dispersal rate of Cane Toads
spinal arthritis, 67–68
Spinifex grass, 216
spondylosis, 67–68
Spotted-tail Quoll, 106
spread of Cane Toads through Australia, 24, 57–78, 59*map*
Starlings, 77, 104
statue of Cane Toad, 20
Stewart, Dinah, 13
St. George, 161
Stockholm Syndrome, 229
Stop The Toad Foundation (STTF), 141, 143, 147–148, 152
straight-line dispersal, 58, 64
Strange Country (book), 136, 141
stress hormone, 32, 54
Stringybark Tree, 42
STTF, 141, 143, 147–148, 152
sugarcane, 7, 14, 16–17, 21–23, 34, 82, 119, 169, 226
Sugar Experiment Station, 18, 20*fig.*
suppression pheromone, 197, 198*fig.*, 207, 228; dependence on egg size, 201; impact on breeding behavior, 200; impact on native tadpoles, 200; use for toad control at invasion front, 207
survival of the fittest, 25
Sutherland Shire Council, 162
symposium, 175

tactics to control Cane Toads, 50, 178, 179, 186, 189, 226; integrated approach, 203; target specific life history stages, 185; translate knowledge into outcomes, 203
tadpole communication. *See* alarm pheromone, attractant pheromone, Bufonite, suppression pheromone
tadpole ecology, 50–51; aggregation, 182; aquatic, 185; consume conspecific eggs, 194
tail-luring by Death Adder, 117
Taipan, 34, 126
Tanami Desert, 32
TAP, 151
Taren Point, 161–162, 162*fig.*

Tasmanian Devil, 106
Tasmanian Tiger, 106, 231
taste aversion learning, 108–115, 209–215, 219, 233; Bluetongue Lizards, 215; fishes, 50, 90; Freshwater Crocodiles, 215; goannas, 215–217; Northern Quolls, 50, 211–218. *See also* teacher toads
Tawny Frogmouth, 139
teacher toads, 203, 209–210, 217–228. *See also* buffering the impact of Cane Toads
Team Bufo, 36, 49–61, 75, 79, 86–89, 101, 116, 130, 136, 150–152, 164–167, 170, 177–179, 200, 203, 210, 222, 227–230, 237
Team Bufo founding members: at Middle Point, 41*fig.*; at Parliament House, 219*fig.*
Territory Wildlife Park, 211
thiabendazole, 108, 211, 215
Thomas, Jai, 215
Threat Abatement Plan (TAP), 151, 251
Tiger Salamander, 85
time course of invader impact, 115, 232, 242
Tingley, Reid, 223, 225
Toad Army, 56, 60, 160, 210, 226, 234
Toad Hall, 63
toad-busters, 1, 10–11, 49, 53, 60–61, 92, 104, 114, 123, 136–139, 140*fig.*, 142, 147, 204, 229, 235; effectiveness, 151. *See also* Frogwatch NT; Kimberley Toad Busters; Stop The Toad Foundation
Toadinator, 159
Toadjus, 160
toad musters, 7, 124
toad politics, 12, 17–24, 134, 143–145, 151–154, 176–179, 210, 237–238
toad sausage, 215–218
toad-smart genes, 112–115, 126, 208, 233
toad toxin causes heart attack, 33
toe-luring by Cane Toads, 184
tolerance of Cane Toads to abiotic extremes: of cold, 30–32; of dry conditions, 32; of salinity, 14
Tongue worms, 87–88
Townsville, Queensland, 114–115, 120, 123

toxicity of toad eggs to fish, 89
toxin, snake tolerance to, 37
toxin glands, 8–10, 196, 204
toxin similarity, Cane Toad vs. plant, 122
toxins of Cane Toads, 9, 37, 128, 195
toxins vs. invasion success, 8
translocation of Cane Toads, 29–30; hitch-hiking on trucks, 28
trap design competition, 158–159
trapping adult Cane Toads, 142, 157–160; white lights repel toads, 159
trapping Cane Toad tadpoles, 193–196; 196*fig.*, 204, 207; commercialization, 207; target specific sites, 187; uptake by community groups, 207
trophic cascade, 118, 232
Tropical Ecology Research Facility, 38, 62–63
tropical mammal decline, 239
tumor cells, 77–78, 106
Turvey, Nigel, 16
Tyler, Mike, 34
Tzu, Sun, 178

Ujvari, Bea, 125
understanding the enemy, 242
University of Queensland, 91, 192, 235
urban heat island effect, 31
use of disturbed habitats by Cane Toads, 15, 28–29, 31
use of media to increase public acceptance of teacher toads, 218–219
using Cane Toads to control Cane Toads, 228
using native wildlife to control Cane Toads, 204
UV light and acoustic cues in toad traps, 159

vacant niche, 93
van Buerden, Eric, 57
Vanuatu, 234
varanid lizards, 119–120, 217*fig.*, 220. *See also* goanna
Venezuela, 57, 165–166
Victoria River, 98, 159
vines, 77

viruses, 38, 83, 165–167, 179
voles, 77
volunteer efforts to eradicate toads, 148, 156. *See also* toad-busting

Wake, David, 76
Ward-Fear, Georgia: research on ants, 50, 188, 204; on teacher toads and goannas, 216–220, 217*fig.*, 225
Wassersug, Richard, 109
water beetles, 128–129, 188
Water Buffalo, 38, 41–42, 47*fig.*, 48, 55, 195, 231, 241
water bugs, 128, 188, 226
Water Dragon, 101
Water Hyacinth, 20
Water Python, 38, 44–46, 52, 102, 119–120, 158, 241
Webb, Grahame, 95, 97
Webb, Jonno, 50, 109–110, 210–214
Western Australia, 25–26, 79, 101, 121, 136, 140–144, 158, 172–173, 201, 207, 215, 217–219, 222, 224–225, 237
Western Australia invaded by Cane Toads, 25

Western Australian Department of Parks and Wildlife (DPaW), 172, 217–218, 237, 251
Western Australian government, 144, 148, 150, 216, 219, 226
wet tropics, 26
Whistling Kite, 43, 105
whitefella goannas, 220
White-tailed Water Rat, 46, 126
wildlife, vulnerability to Cane Toads. *See* poisoning of wildlife
Willow Tree, 55
Woinarski, John, 91
Woolly-butt Tree, 42

xenophobia, 135, 229

year-round activity of Cane Toads, 28, 99, 114
Yellow-spotted Goanna, 28, 99, 111, 114–115, 217*fig.*, 221–222, 221*fig.*, of cane toads, 233; impacts of indigenous hunting, 232

zombies, 9

ABOUT THE AUTHOR

Rick Shine is Professor of Biology at the University of Sydney. He has published more than a thousand scientific papers on the ecology of reptiles and amphibians, and he has received a host of national and international awards for his research.

ORGANISMS AND ENVIRONMENTS

Harry W. Greene, Consulting Editor

1. *The View from Bald Hill: Thirty Years in an Arizona Grassland,* by Carl E. Bock and Jane H. Bock
2. *Tupai: A Field Study of Bornean Treeshrews,* by Louise H. Emmons
3. *Singing the Turtles to Sea: The Comcáac (Seri) Art and Science of Reptiles,* by Gary Paul Nabhan
4. *Amphibians and Reptiles of Baja California, Including Its Pacific Islands and the Islands in the Sea of Cortés,* by L. Lee Grismer
5. *Lizards: Windows to the Evolution of Diversity,* by Eric R. Pianka and Laurie J. Vitt
6. *American Bison: A Natural History,* by Dale F. Lott
7. *A Bat Man in the Tropics: Chasing El Duende,* by Theodore H. Fleming
8. *Twilight of the Mammoths: Ice Age Extinctions and the Rewilding of America,* by Paul S. Martin
9. *Biology of Gila Monsters and Beaded Lizards,* by Daniel D. Beck
10. *Lizards in the Evolutionary Tree,* by Jonathan B. Losos
11. *Grass: In Search of Human Habitat,* by Joe C. Truett
12. *Evolution's Wedge: Competition and the Origins of Diversity,* by David W. Pfennig and Karin S. Pfennig
13. *A Sea of Glass: Searching for the Blaschkas' Fragile Legacy in an Ocean at Risk,* by Drew Harvell
14. *Serendipity: An Ecologist's Quest to Understand Nature,* by James A. Estes
15. *Cane Toad Wars,* by Rick Shine